Biogas For Beginners

George Eccleston

George Eccleston

Copyright©2018 George Eccleston

Produced in the United Kingdom

All rights reserved. No part of this publication may be reproduced, stored in a retrieval system, or transmitted in any form or by any means, electronic, mechanical, photocopying, recording or otherwise without written permission of the author or publisher.

ISBN: 9781724142221

Imprint: Independently published

Important warning

Biogas creation does come with some inherent risks associated with it. Biogas is very flammable & explosive when mixed with oxygen, so great care must be taken when using or creating this kind of gas. Inside the biogas are toxic gases such as hydrogen sulphide & Carbon dioxide gases to name just two, These gases must be dealt with & controlled in a suitable & safe manner.

Although care has been taken to inform you of these risks and suitable safety precautions have been included within this book, the building of biodigesters and the creation of the gases from this apparatus does entail some degree of risk and it is solely your responsibility to manage and deal with these risks.

Disclaimer

Although the author and publisher have made every effort to ensure that the information in this book was correct at the time of publication, the author and publisher do not assume and hereby disclaim any liability to any party for any loss, illness, damage, or disruption caused by errors or omissions, whether such errors or omissions result from negligence, accident, or any other cause.

Table of contents

Chapter 1	Biogas for beginners	1
Chapter 2	Bacteria and methane production	4
Chapter 3	Different types of biodigesters	8
Chapter 4	Urban biodigesters	14
Chapter 5	Basic safety devices	25
Chapter 6	Storing biogas	32
Chapter 7	Cleaning dirty gases	36
Chapter 8	Bottling the gas	40
Chapter 9	Keeping them warm	47
Chapter 10	Feeding the digester	51
Chapter 11	Checking the PH levels	55
Chapter 12	Converting gas stoves to use biogas	58
Chapter 13	Dangers & precautions	61
Chapter 14	Finding the prime suspect	63
Chapter 15	More books by this author	65

Chapter 1

Biogas for beginners

Welcome to the wonderful world of biogas production. Biogas in case you are unaware is a type of bio-fuel which is created using bacteria and waste products such as manure, food scraps, vegetable cuttings and many other type of waste organic matter.

You can use biogas in the same way that you would use LPG (Liquefied petroleum gas) but the difference is that you can make biogas at home free of charge using waste food scraps. This free gas is created by bacteria called (archaea) inside an anaerobic (air free) environment. These bacteria digest the waste products and give off methane gas as a by product. It is this methane gas by-product that we use as a burnable gas. We house the bacteria inside an airtight container where they can digest these food scraps, this is why it is called a **biodigester** (short for a biological digester), and sometimes it will be shortened to simply a **digester** (it is referred to as both throughout this book).

Usually when you throw away food scraps they get taken away by the garbage men and buried underground where there is no oxygen. This is an anaerobic environment which means that this food will be digested by the methane creating bacteria underground. All the methane that they create will eventually find its way up to the surface and is released into the atmosphere. Methane gas is over 25 times more potent a greenhouse gas than CO2. By throwing your food out into a standard garbage bin you are contributing to global warming possibly more than any other thing that you are doing.

You may have a food collection and recovery service where you live and if they are good at what they do then they can convert this into compost or biogas which is great. The big question is why you would want them to do this when you could do this yourself and get free cooking and heating gas which will reduce your utility bills.

Nature has its own method for creating methane gas. Animals and humans have bacteria living inside their gut that help to breakdown the food that they eat. A by product of the bacteria digesting food is the fact that they give off methane gas.

Cows are exceptionally good (or bad depending on your point of view) at producing methane gas. The food goes into the cow's mouth and is ground up by the teeth. Then it is passed down to the stomach and intestines to be digested.

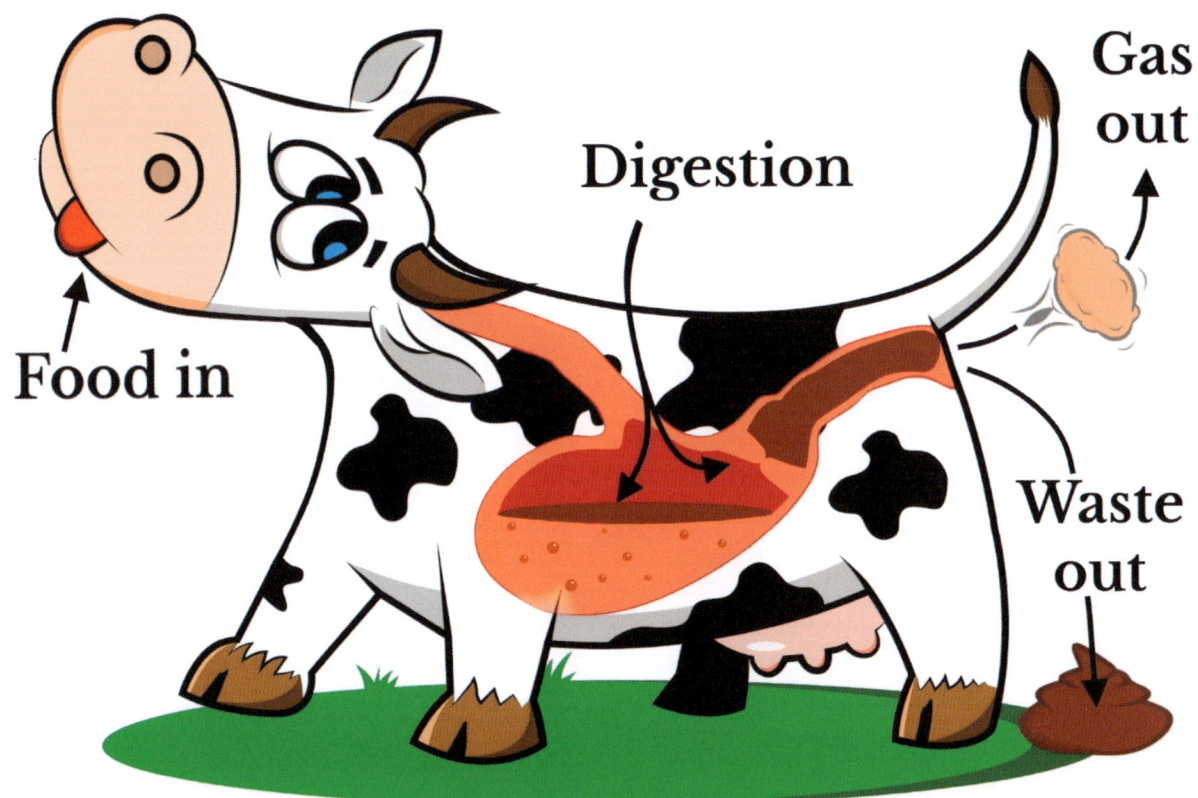

Although this picture is not anatomically correct it does give you an idea of what is going on inside the cow. The bacteria that live inside the animals gut consume some of this food and then excrete chemicals, amino acids and hormones that are essential for the cows (and humans) health and wellbeing. The bacteria also give off methane gas as a by product. Waste products in the shape of manure are excreted from the back end of the cow. The methane gas that is created inside the cow is also released from the cow's backend at the same time. The problem is this methane is released into the atmosphere and wasted. The escaping gas also causes great environmental problems because methane is a potent greenhouse gas.

We can recreate and actually improve on the way nature produces methane gas by building ourselves a biodigester. The biodigester can be made very easily and quite cheaply using nothing more than plastic water tanks or barrels, hose pipe, plastic pipes and a cheap valve or two.

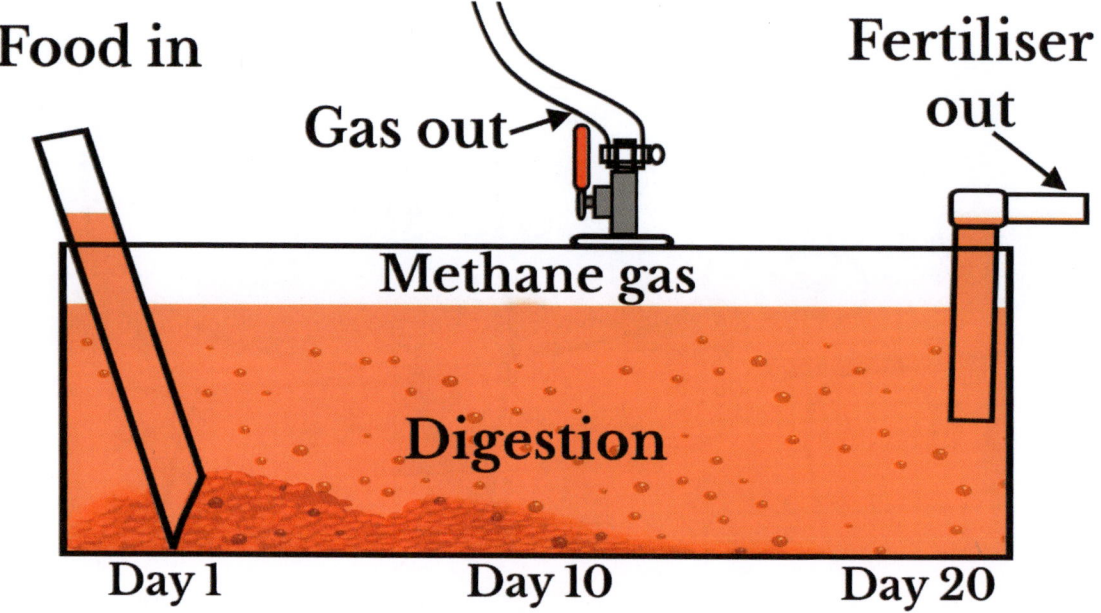

Although this doesn't look very much like the cow it does still have a mouth where the food gets put into it and it also has a stomach where the bacteria digest the food. The methane gas rises to the top where it can be extracted via the hose on the top. The digester also has a waste disposal system like the cows backend but the difference is that this waste product is high quality liquid fertiliser that can be placed straight onto your garden.

The simplified diagram shows one possible type of biodigester there are various designs to choose from as you see later in the book.

Chapter 2

Bacteria and methane production

The bacteria that create the methane gas for us are called Archaea bacteria. They need to live in an oxygen free environment. Microorganisms that can produce methane gas are called Methanogens. There is some dispute as to whether the Archaea should be classed as bacteria (no cell nucleus) or Eukarya (with a cell nucleus), but for this book they will just be referred to as bacteria for the sake of simplicity.

There is a large variety of anaerobic bacteria which exist all around the world. They have been on this planet long before our human ancestors came into existence and they will be here long after we are all gone. They are one of the planets hardiest species that can exist and thrive in most types of environmental extremes anywhere on the planet. They can even survive in volcanic springs in high temperature areas that nothing else seems to be capable of surviving in. Like superman they do have one weakness but it isn't kryptonite, it is oxygen. So long as they can keep away from oxygen they will be happy and productive members of society that will happily produce free methane gas for you.

Oxygen is actually toxic to these bacteria which is why they hide in places where there is little to no oxygen such as under the ground, under water and inside the gut of animals.

Choosing the correct types of bacteria

Because there are many different types of bacteria that can be used it helps you if you narrow down which particular bacteria that you can use for producing methane gas.

Psychroptilic bacteria produce gas at low temperatures, these can live very comfortably in cooler climate. Psychrophile which means cold loving in the greek language describes these bacteria perfectly because they can survive in cold temperatures down to minus 20°C.

Mesolphilic bacteria prefer to produce methane gas at temperatures similar to which the human body prefers, this is around 37 degree Celsius (98.6 Fahrenheit). This means that they are happy to be in unheated containers in your garden in the summer months. Generally speaking if you are warm enough then so are they. This type of bacteria may not produce quite as much gas as the Thermophilic bacteria but they are so much easier to care for.

Thermophilic bacteria demands quite hot temperatures to live in and they won't produce much gas unless you meet their demands for a nice warm home. This type of bacteria does tend to produce more methane but they can be very fussy and easily upset if you don't give them the correct type of food or they are not kept at a constant temperature.

Finding these bacteria

Which bacteria are best for you is generally dictated by which part of the world that you live in. Generally speaking the Mesolphilic bacteria are the favoured type because they favour the environments that we do ourselves, if we are warm enough so are they. In the winter however when you are in a nice warm building they are sat out in a freezing cold back garden so they will have problems producing gas in these conditions unless you treat them like a human and give them a nice warm jacket (of insulation) and warm them up somehow. This is quite possible and actually fairly easy to do but this will be covered later in the book.

Finding the bacteria that you will need is actually a very easy task because as we already know they live inside the digestive tract of animals. This means that every time an animal leaves behind a deposit of its faeces on the ground, it also leaves an array of the bacteria that we need stored inside these faeces. For those people that are a bit squeamish we shall call these faeces manure from this point on. We want to collect the manure as fresh as possible so that less of the bacteria have a chance to die.

How do they convert food into gas?

Microbes that use oxygen will attack and consume a fallen fruit and eventually consume it but it may take a month or two for it to fully decompose. And the gas that these microbes give off would be useless to us because it is CO_2 (carbon dioxide), which will not burn.

Oxygen breathing microbes are very slow at eating in comparison to the way the archaea will digest this fruit. The archaea will attack and convert this fruit literally overnight and within a few days there will be nothing recognisable about this fruit left. You do have to help them along by grinding or mashing the fruit into a pulp similar to the way an animal chews its food up before it eats it, but even so they are so much faster at converting this organic matter into a gas than their oxygen breathing counterparts.

How does a biodigester work?

We shall go into more detail later but the basics of how methane gas is created inside a biodigester are very easy to grasp. The three main ingredients required to make methane are fresh manure that will be rich in bacteria, water and an airtight container or water tank for the bacteria to live in.

Manure   **Water**   **Airtight Container**

Once we have obtained the manure that will be rich with bacteria we mix it with the water and then place it either inside an airtight container or keep it under water inside a water tank where the air cannot get to it. We then allow the bacteria to feed and multiply. In return for feeding the bacteria on a regular basis with either manure or the food scraps that we don't want, the bacteria will give us their waste products (methane gas). This is the basic cycle that needs to happen although there are a few steps in-between that must take place to ensure that this process happens the way we want it to.

Bacteria convert food waste into methane using a process called anaerobic digestion. There are a series of steps that microorganisms use to breakdown biodegradable materials in an oxygen free environment. The main steps are listed below.

The four main stages to anaerobic digestion

First of all you need to understand that there isn't just one type of bacteria breaking this food source down into gas, it is a collaborative effort by different types of bacteria and various chemical reactions. Each type of bacteria eats the food that it wants and creates a very specific type of by product that can be used by the other species of bacteria, which eventually leads to the final desired result of gas production. This is a four stage process and each step must take place in order for the final desired gas output to happen.

Hydrolysis- This is the process were water is mixed with the food and converts particles of solid foods into a liquid form. The complicated polymers that hold the food together are broken down into simpler structures which make it easier for the microorganisms to convert it into their food. In technical terms this means that the long complicated structures inside the food (Polymers) are broken down into smaller easier to handle portions (Monomers). Processing the food yourself by grinding it up (like the cow chewing the food) into very small pieces and mixing it with water will not only make it easier for the bacteria to eat this food but it will also make the gas production happen much faster because the bacteria don't have to waste time attacking the large chunks of food and slowly breaking them down.

Acidogenesis- It's at this point that the bacteria start to ferment the food and the simpler structures (the monomers) are converted into volatile fatty acids.

Acetogenesis- At this stage the bacteria convert the volatile fatty acids into acetates (mild vinegar type acid), carbon dioxide and hydrogen gas.

Methanogenesis- At this stage the acetates that have just been produced can now be converted into methane gas. Once all the acetates are consumed hydrogen and some of the carbon dioxide can also be used and converted to methane gas as well.

If you take care of the biodigester and the bacteria inside it then you get an ample supply of methane gas produced by these contented bacteria. We shall go over the correct ways of caring for the bacteria later on but first we shall look into building the bacteria their new biodigester home for them to live in.

Chapter 3

Different types of biodigesters

It is good to know how the bacteria function and create the gas but what probably interests you more is how you get the bacteria to create gas for you and how can you capture this gas. This is actually quite simple because all you have to do is create a good comfortable home for the bacteria to live in and remember to feed them every day. Then you can simply siphon off the methane gas that they create. These homes are called biodigesters and they can be quite simple to construct, even if you don't have any DIY skills.

There are various types of biodigester designs from all around the world that you can choose to build from. Each different design has benefits and drawbacks so it is up to you to decide which design is best for your needs. We shall go through a few of these different designs and show how they are constructed and used and then we will go through the building and initial setup and maintenance of a biodigester itself. Once you understand the basic principles involved in building one type of biodigester you should be able to build most types without any further help.

To increase your knowledge base of different types of biodigesters we will cover some of the more complicated industrial sized biodigesters first before we move onto the simpler more practical urban designs.

Chinese fixed dome

Building a Chinese fixed dome digester is a big undertaking demanding a lot of time, labour and cash. This is because it is built underground with a concrete base, brick walls and a concrete dome to hold the gas inside it. The waste products are fed into the digester above ground and broken up using the mixing paddle. A plug is then released allowing the waste to run through the inlet pipe into the digester. It then slowly digests giving off methane gas that will be trapped underneath the concrete dome.

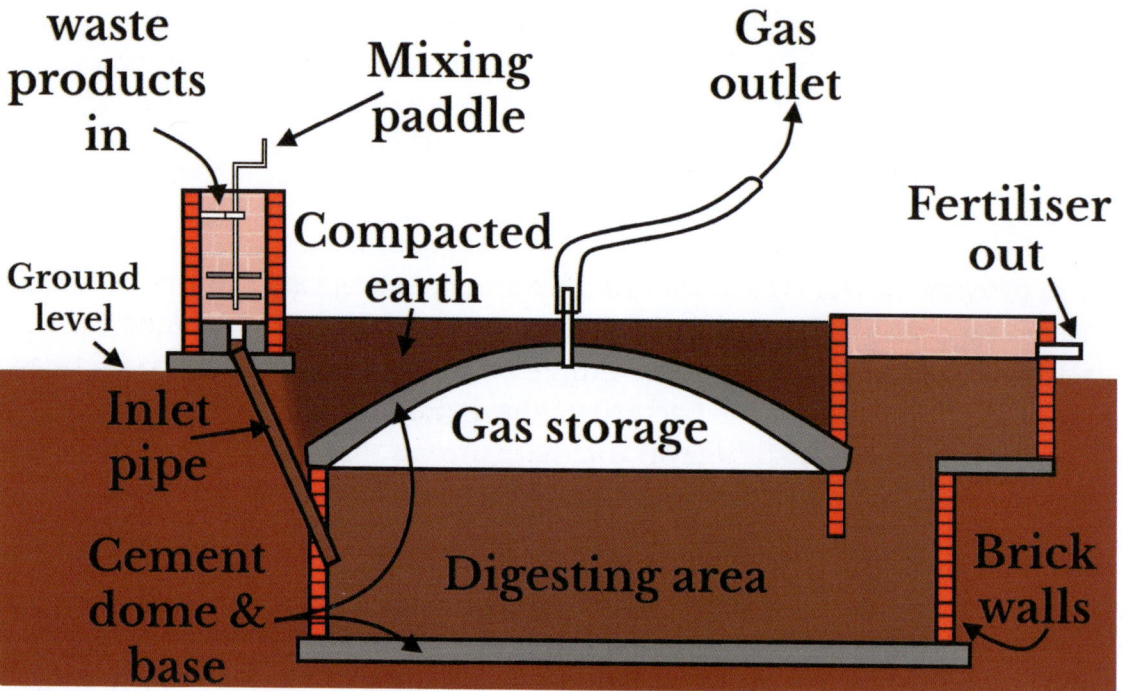

When new feedstock is added the liquid in it displaces some of the liquid at the far edge of the digester forcing it up and out of the biodigesters fertiliser outlet pipe. This fertiliser can be used on vegetables and other plants as a fantastic plant food. Because the incoming liquid & waste products are above the digester itself the weight of the water pressurises the gas inside the dome. The gas is released from the pipe that goes through the top of the dome and is piped off to where it needs to be used.

Used primarily for digesting a lot of animal waste these are frequently used on farms across Asia.

Complete mix tank reactor (CSTR)

With this type of biodigester the manure is fed into one side and pulled out the other. As it passes through the reactor it is thoroughly mixed by a paddle. Floating within this mixture are sand particles that the microbes are living on. These are suspended in the liquid allowing the microbes to get to their food. The temperature is constant all throughout the mixture because it is being mixed together all the time. These are usually built with heating pipes built into the design to obtain the very best temperature for optimum gas production. It can be fitted with a flexible membrane cover that expands when gas is produced and holds the gas above the tank, or it can have a solid rigid cover with the gas stored elsewhere.

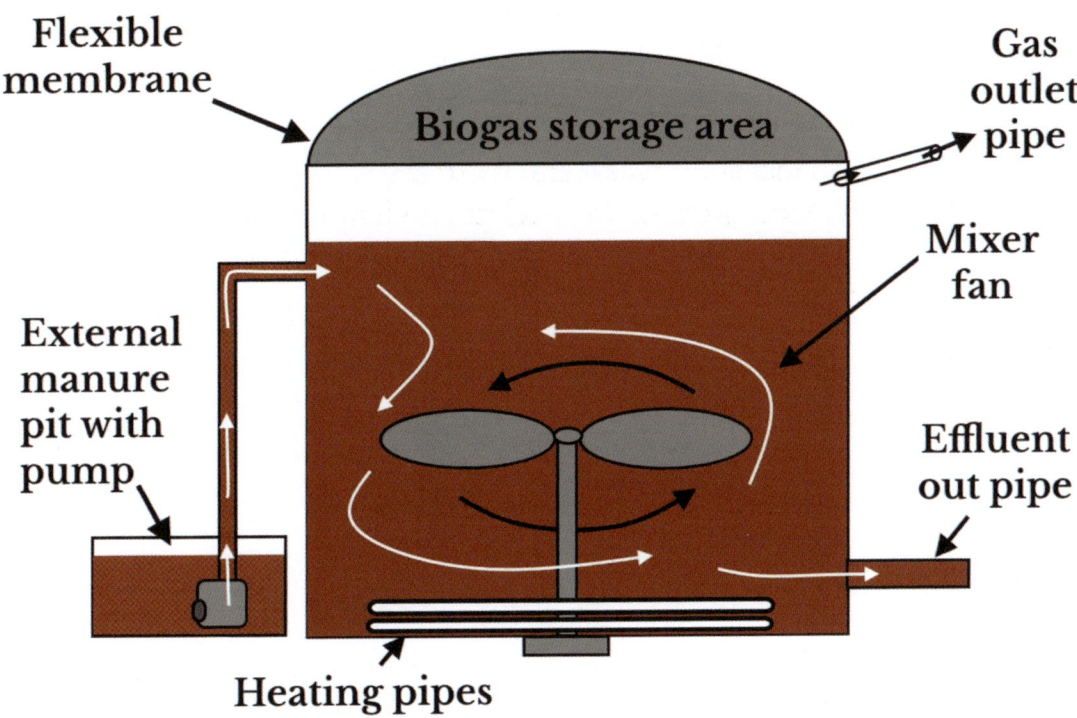

The problem is that the feedstock that you have just put into the tank can easily leave the tank a few minutes later because the liquid passes through the tank in an uncontrolled manner. Generally speaking the manure is in there long enough to produce a high quality gas but sometimes it does come back out of the tank long before it has been processed properly.

Covered lagoon reactor

The covered lagoon design is created by having a lagoon shaped hole sunk into the ground. This pit is then filled with manure and covered with a membrane to trap any gas that comes off it. The pit must be watertight & airtight to contain the effluent and to stop it leaching into the ground water below. Most of the lagoon type digesters are built with a concrete tank to hold the effluent and then covered with a gastight membrane.

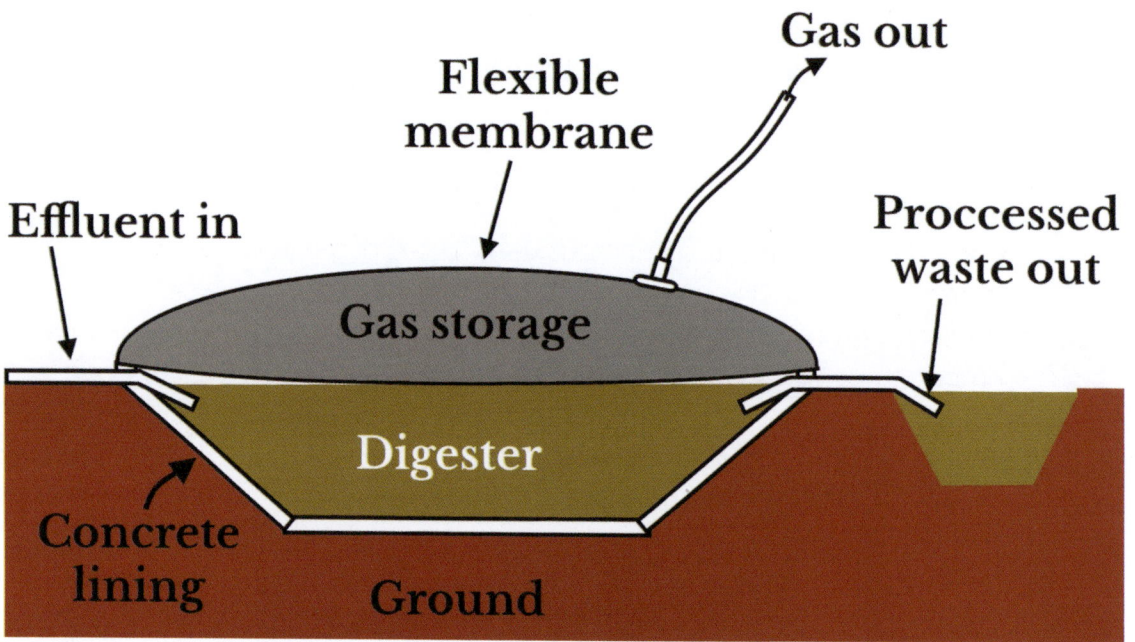

This is a popular design because of its low cost construction. Its large surface area is heated by the sun and methane soon starts to be produced.

You can get smaller versions of this digester specifically for your back garden which is about the size of a large paddling pool. These are built above the ground making it easy to install. These are easily ordered online and are made from a sturdy plastic. They can work reasonably well but they are very expensive to buy. Meaning that you could build half a dozen of your DIY versions for the same amount of cash, if not less.

The lagoon digester doesn't usually require stirring of the mix inside so that is one added benefit but it is not the most efficient producer of gas and it does take up quite a bit of space.

The flexible Membrane that this design uses can actually be used on various types of biodigesters. It is a very cheap and easy way to create gas storage using low cost pond liners or rubber membranes.

Indian floating dome

This is called the Indian floating dome because it was the favoured design for many years across India. It is primarily used in farming areas in the countryside but can also be found anywhere that animal waste is plentiful. Primarily fed manure it does need to be quite large in size to handle the amount of animal waste that is produced everyday on farms. It is an ideal solution for these farmers because the manure which is a health hazard and a waste product can easily be processed in the digester giving valuable fertiliser for the crops and gas for cooking food & heating.

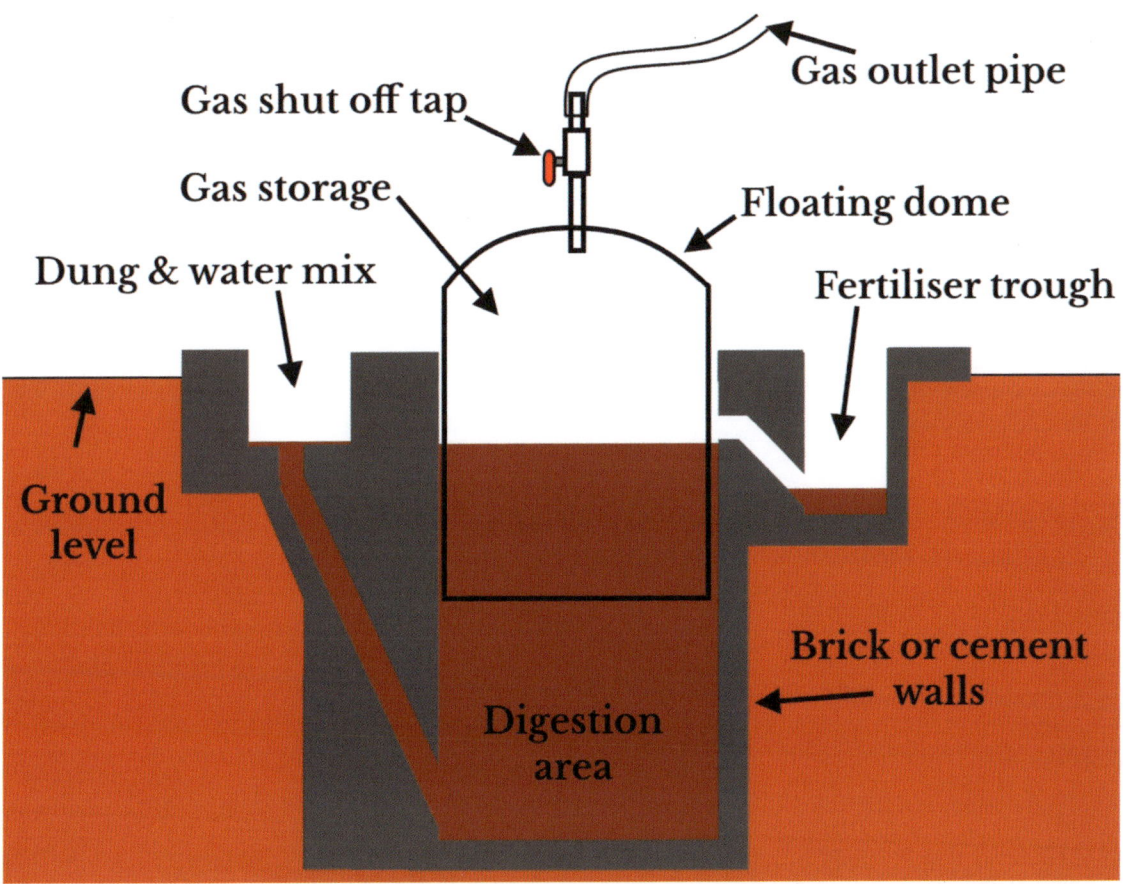

An upside down metal drum (dome) floats on top of the manure & water mixture capturing any methane that is produced by the bacteria. The more gas that has been produced the higher the metal drum will float upwards. This simple technology with very few moving parts is easy to maintain and lasts for quite some time.

Generally speaking these digesters don't produce massive amounts of gas because they are fuelled using manure which is reasonably low in calorific value for the bacteria. But what the manure lacks for in energy it makes up for in volume. Because so much manure is usually produced on a farm the farmers generally speaking have plenty of usable gas for their own needs, so it is a great system for them.

These designs are all large scale digesters for use with large quantities of manure. For use in standard size houses and gardens these digesters are far too large. In the next chapter we shall cover digesters designed for urban dwellers that don't have access to an excess of free manure to feed the bacteria with.

Chapter 4

Urban Biodigesters

You have a lot of options over which types of materials and various different shaped containers to build your urban biodigester out of. Before you start to build any of these you will first need to know one important warning about any container or water tanks that you choose to use. The containers (barrel, water tank, bowser tanker, plastic tubs etc) that you use to construct your biodigester from cannot allow sunlight through to the inside in any way shape or form. This is because this would encourage the growth of algae which produce oxygen as a by product, and oxygen will kill your methane producing bacteria stone dead. It is for this reason that you should never ever use transparent containers and tanks unless you paint them black so that they totally block out the light preventing any algae formation. The black colour will also have the added benefit of absorbing the heat from the sunlight which will heat up the inside of the digester nicely.

The designs that we shall cover in this chapter are ones that can easily fit into an urban back garden because they demand very little space and can produce a surprising amount of methane gas for you to use. We will start with the most basic & primitive design and work our way up from there. We will only cover the design of the digester for now, as far as storing the gas is concerned we will cover that in the following chapter.

The two barrel batch system

The two barrel system is just as it sounds, you simply use two separate plastic or metal barrels to provide the gas that is needed for your cooking needs. This is what's known as a batch converter, because it processes all the food inside the barrel in one go rather than processing small amounts over a period of time. This makes this design ideal for people who are unlikely to remember to feed the digester regularly.

This setup relies on you having access to a fresh supply of manure roughly every 5 to 6 weeks. This is because you are only going to use one barrel at a time and rotate the use of these two barrels so that you always have a constant supply of gas.

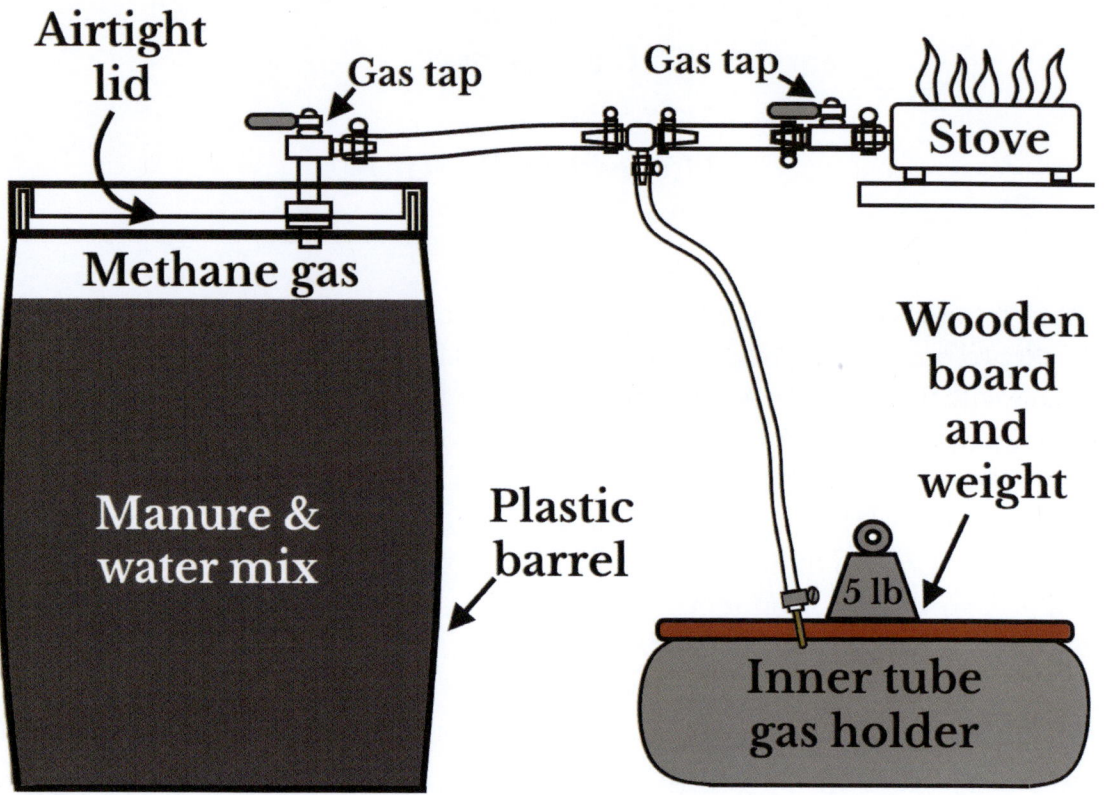

This diagram is shown with an inner tube gas storage device, various gas storage devices will be covered later on in the book.

This barrel is half filled with manure and then it is topped off with water till it is nearly full. You will have to leave a couple of inches room from the top because the mix expands once it starts to ferment and you don't want this mix clogging up the gas outlet pipe at the top of the barrel.

The first week or two will produce mainly CO_2 (carbon dioxide) and other useless gases which simply will not burn properly. This means that for the first week or two at least it is a good idea to put the hose that leads to the gas storage devices into a deep bucket of water and allow the gases to simply bubble up through the water. With the end of the pipe underwater it means that air cannot get back into the barrel whilst the remaining air and CO_2 gases inside the digester will be forced out of the barrel as more & more gas is produced by the bacteria. After about two weeks the hose can be reconnected to the gas storage device properly (not forgetting to close the gas valve before it is removed from the bucket of water).

Removing any remaining oxygen in this manner from the barrel will make it absolutely safe because methane cannot catch fire or cause explosions unless there is oxygen mixed in with it. Getting rid of the oxygen also does away with any smell because only when it is mixed with oxygen does the manure smell.

You will only open this barrel once in order to fill it with the manure and water mix. It isn't opened again until all the methane has been extracted from the barrel in about two months time. You don't have to feed it anymore food or manure during this time because if you did air would get into the barrel killing lots of the bacteria and releasing the gas. It is basically a fill & forget barrel. All you have to do after it has been filled is to remember to use the gas to cook your food with.

It will take the bacteria at least two to three weeks to start producing the methane gas. Once it does you should have a regular supply of methane gas for cooking.

One large 200 litre drum should carry on supplying gas for around two months before the food source for the microbes inside it is all gone. The big secret of using drums to provide a constant supply of gas is to plan ahead. After one barrel has been providing gas for about 5 weeks you will need to set another barrel up with fresh manure & water because this will take about two or three weeks to start providing usable gas. This means that by the time that one barrel has come to the end of its useful life the other barrel is ready to take over providing the gas for your needs from that point on.

This is an ideal system for those that don't want to be bothered chopping or grinding food waste everyday in order to feed the digester. There is much less work to this system and it happily sits there providing gas for your cooking needs. This should hopefully give you about 40 minutes worth of cooking time per day during the summer months.

Once the gas has all been extracted you can use what is left inside the barrel as liquid fertiliser on your garden. It will look and smell a lot better than when it went into the barrel and it is still very nutritious for your garden plants. If it is opened before it is fully digested and finished with then you will find that the liquid inside can be particularly smelly and disgusting so ensure that it has used all of its gas up first.

This is a cheap and quick method of producing biogas with very low setup costs. It takes up very little space so it is a very good starting setup for the beginner.

All the following digesters are classed as continuous process digesters because the food is added every day rather than just in one big hit.

Solar cities IBC biodigester

This next type of digester could be of great use to you. It has been designed by Solar Cities which is an open source platform designed to encourage as many people as possible to build their own biodigesters. You can find their website online at http://www.solarcities.eu/education/388 where you can download the plans for an IBC DIY build biodigester free of charge.

In the Solar cities design they use old IBC (Intermediate Bulk Container) tanks because they are cheap and plentiful. The idea to build it from old recycled IBC tanks means that most people can afford one. The fact that they hold 1000 litres of water is also a fantastic bonus. The beauty of this design is that it can be adapted to suit any reasonably sized water container that you can get your hands on. I personally struggled to get a second hand IBC tank because it simply wouldn't fit into the vehicle that I have and most of the places that sell the recycled IBC tanks seem to have a strict no delivery rule.

I settled on a slightly smaller design and chose a different container that was much easier to come across and carry in my vehicle. Using a little bit of lateral thinking can get you close enough to the solar cities design to create a fully functional biodigester from what you can get your hands on. In this case my water container was actually a 240 litre capacity wheelie bin. The lid section on the bin needs to be sealed so it is 100% airtight but once this is done it is good to go. They take up very little space and are made from virtually indestructible plastic so it was ideal for my needs. It won't hold as much liquid as an IBC tank but it isn't bad.

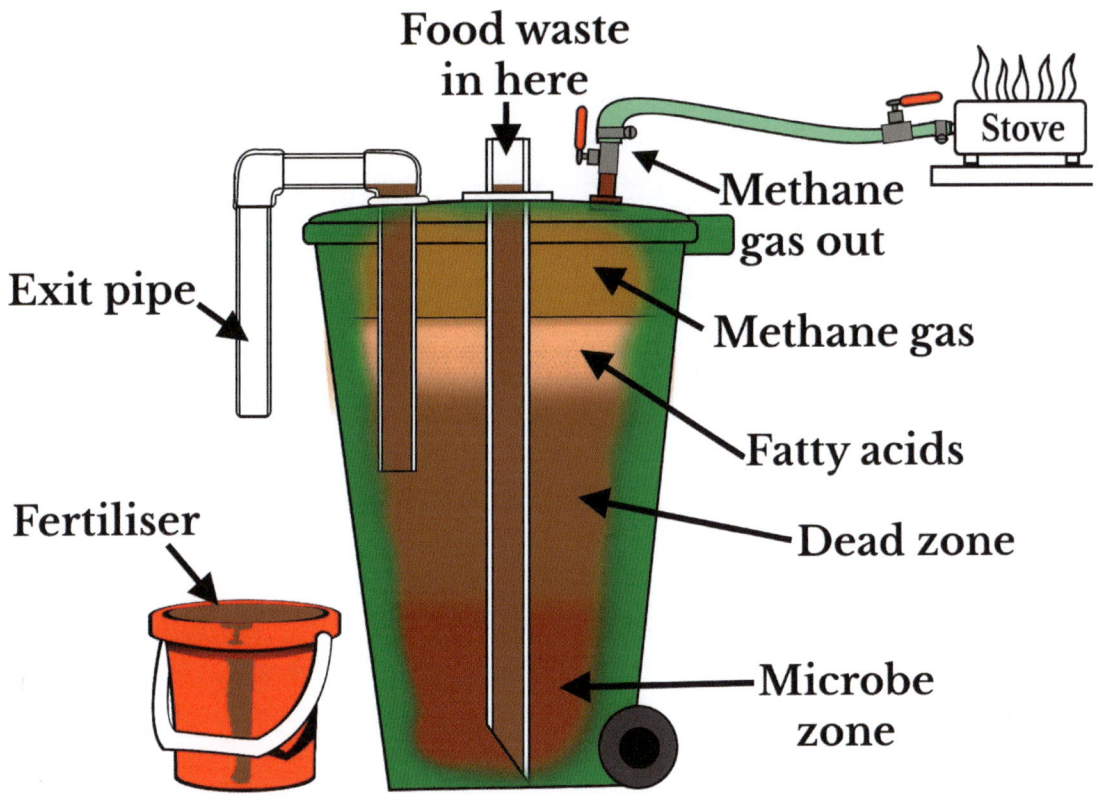

This diagram is shown without any gas storage for simplicity only.

Now before you go using the wheelie bin provided by your council don't forget that these are officially the property of the council unless you have actually bought it off them or another company that provides them. Keep it legal by purchasing your wheelie bin as they are not overly expensive and they get delivered to your door, how easy is that?

The ground up food goes into the central inlet pipe and goes straight to the bottom of the bin where the microbes begin to process the food immediately.

There are three distinct zones in the container. There is a microbe zone which is where most of the bacteria live, the dead zone in the centre were very little happens or lives and the surface of the liquid where the fatty acids that tend to float upwards linger and get converted into more useful by products.

When you extract the excess liquid fertiliser the last thing that you want to do is to accidentally take to many bacteria out of the biodigester whilst you are doing this. This is why it makes sense to take the fertiliser from the middle dead zone section where there are very few bacteria. This is why the exit pipe is designed to extract the fertiliser from the centre of the container where this dead zone is.

Methane gas is trapped under the lid of the bin and can be siphoned off as needed or sent to a gas storage device (covered later) for later use.

Plug flow reactor

This next plug flow digester does take up a bit more room but it can still be fitted into a small back garden quite easily. It is a horizontal digester so it is quite a bit longer but this isn't a huge problem as it can be run alongside a fence so it is out of the way. You can even place a bench top over it so that it can be used as a seating area (or possibly an outside table/kitchen area). This is a great way of disguising it and allows you to wrap it with insulation for the winter (more about insulation later).

This design is commonly used in large scale commercial digesters but it can be easily scaled down to suit smaller everyday household requirements. This is a fairly low cost design which is easy to construct. It can be made quite simply from a very large diameter plastic pipe (the wider the better) with watertight (and airtight) caps on each end. You can also use a large water bowser, the kind that you get fastened to trailers for transporting water around in emergency situations. If you can find a second-hand one it will be much cheaper. Any watertight container that can be run horizontally across the ground will do, so that the contents inside it can travel from one side to the other very slowly.

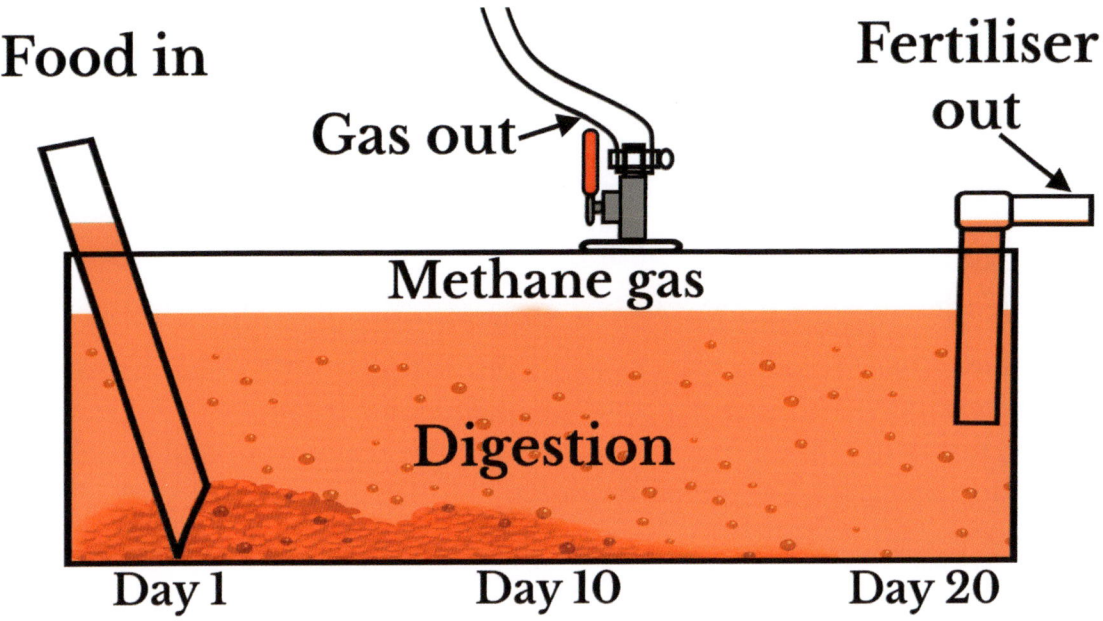

A gas storage device hasn't been shown in this diagram but one should be used.

When new scraps are added they will be in the day 1 position in this diagram. By the time these scraps of food have reached the end of the tank they will have been inside it for 20 days and there will be virtually nothing left of them because the bacteria will have devoured them thoroughly and converted them into methane gas. There is very little maintenance and it gives a good quality smell free fertiliser as a by-product.

The plug flow biodigester will need an inlet and outlet pipe on opposite ends of it. On the top of the digester you fix a hose to siphon off the gas as it is produced

It can also be made by creating a large wooden crate and lining it with a pond liner to hold the water mix inside it. This has to be sealed (100% airtight) on top with more pond liner which could be used to create a flexible storage area for the gas to be stored in depending on what storage option you choose from the chapters yet to come. It can even be buried inside a long trench in the ground so that it is invisible to see apart from the inlet & outlet pipes, but this means hard work and gives you difficulties keeping it warm in the winter unless it is insulated.

This design or one very similar has also been manufactured by many a DIY builder very cheaply in the past by welding two or three steel barrels together to make one extra long metal digester tank at a very low cost.

This design is so good because the waste food cannot leave the tank until it gets all the way to the other side of the digester (which can take weeks depending on how long the digester is) which means that the food scraps should all be thoroughly digested by the bacteria and all the methane extracted from it before it leaves the digester itself, making this quite an efficient system.

It requires no electrical power because all the movement inside the digester is created by either water pressure or gravity.

Some of these have a rotating mixer paddle inside them to turn and mix the contents thoroughly. This can improve the quality and amount of gas produced, but is not a requirement it is an option. This will involve more work and cost but it does boost performance a bit. I personally wouldn't bother but it's up to you.

The ARTI biogas digester

This last digester design combines the basic digester with a built in gas storage device so that it takes up much less space and is very easy to build. It is similar in design to the Indian floating dome digester but it uses plastic barrels instead of concrete & bricks.

The **ARTI** biogas digester (Appropriate Rural Technologies Institute) was primarily created for people living in urban areas of India. Their website is located here http://www.arti-india.org/. These digesters were designed for people who don't have access to livestock waste for converting into biogas but do still have plenty of household organic waste that could be utilised to make gas with. It was developed to use vegetable waste, fruit peelings, meat waste and anything that contains sugar, protein and starches.

It was specially designed to take up as little space as possible because urban dwellers tend to have less space than their rural counterparts. So long as people have a small space on their balcony or in a corner of their garden they can usually fit one of these types of digesters into it somewhere. With its small footprint many more people have found themselves able to produce their own gas and dispose of their household waste in an eco friendly manner with the added benefit of free fertiliser.

It is constructed using two HDPE (high density polyethylene) water tanks. One water tank has to be slightly smaller than the other one so that it can slide inside the bigger tank like in the diagram below.

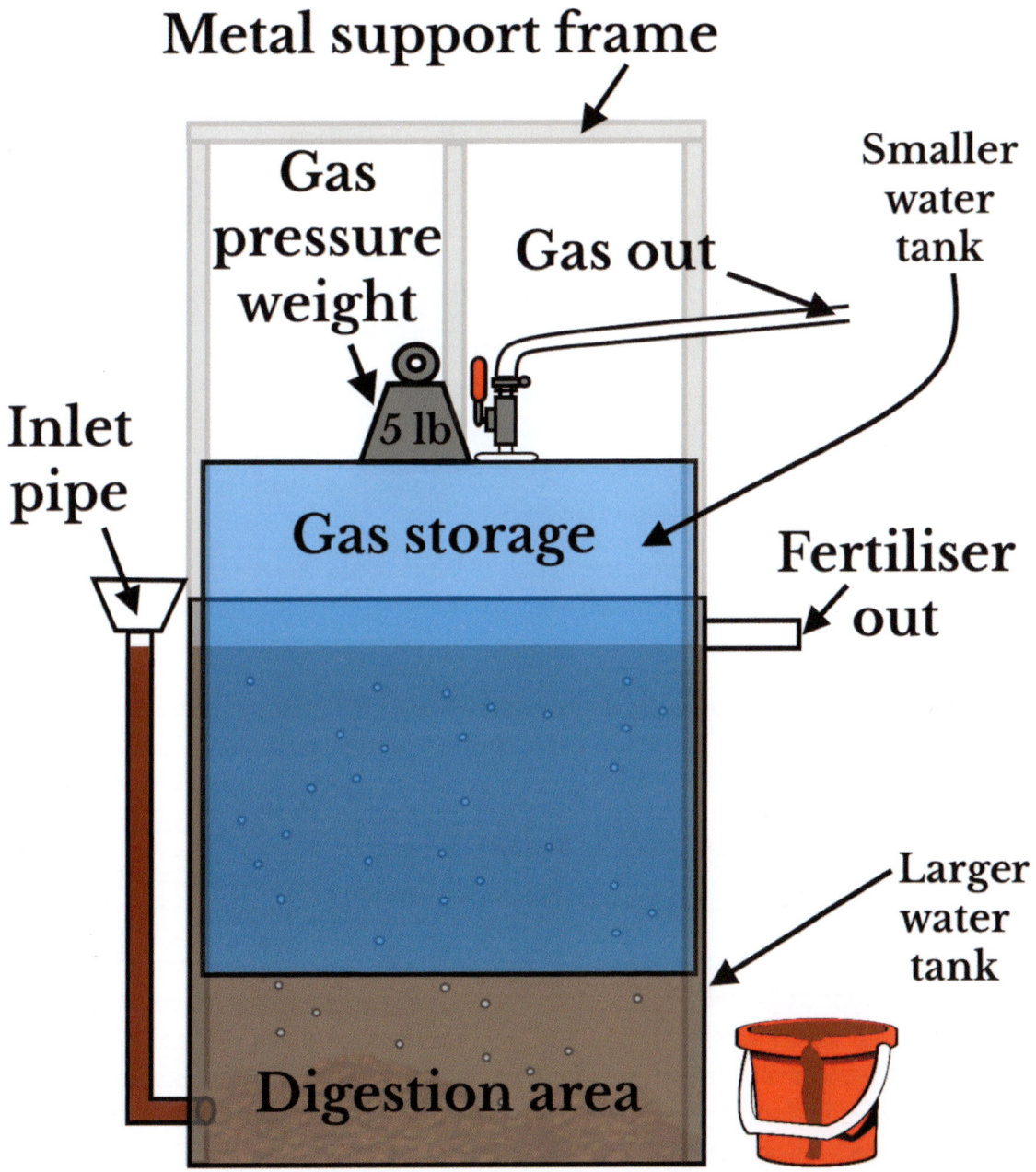

The larger bottom plastic tank sits on the floor and has the top of it cut away completely. This larger bottom tank is the digester section (the stomach of this device) were all the food will be digested by the bacteria. Food will be tipped into the bottom of this tank via the inlet pipe. It will then be eaten by the bacteria who will give off methane as a by product.

The best thing about this biodigester is that it has its own gas storage area built into the design. The smaller tank which has also had its top removed has been flipped upside down and pushed inside the bigger tank. This smaller tank captures most of the methane that is produced underneath it. When new methane gas is trapped inside this smaller water tank the pressure of the gas inside it will make it rise higher upwards. This tells the user exactly how much gas is trapped inside it and gives them a rough idea of how long a cooking time is left in it. The telescoping action of the top barrel means that there is no need for any other containers or gas storage devices, everything is produced and stored in one small area.

The weight on the top of the smaller tank pushes down on the barrel which increases the pressure of the gas inside it, making the gas much better to cook with. The gas is released from the barrel via a tube which has been attached to the top of the smaller water tank. The built in gas tap allows you to stop and start the gas flow whenever you wish.

The metal support frame prevents the smaller inner tank from tipping over as it rises upwards preventing it from releasing all of the methane out into the atmosphere.

As soon as fresh food & liquid is added to the digester the processed liquid fertiliser at the other end of the tank will leave the system via the fertiliser out pipe, so have a container ready to catch it as it leaves.

As with all these urban design biodigesters these are not designed for use with manure only, it will however need manure to start it off and put the correct types of bacteria inside the digester when it is first built. Once these bacteria have multiplied then they will be more than happy to eat proper food rather than manure (in fact they prefer it). Instead of using animal manure like their rural neighbours the urban folk can get much better gas production using food types that are very high in sugars and starches. This doesn't mean that they are tipping bags of sugar into the digesters it merely means that they are utilising high sugar food waste such as fruit & vegetable peelings, waste pastries and other types of high sugar, fat or protein rich foods. The higher sugar, fat & protein content is loved by the bacteria and they will produce much more gas from this than they would from manure which has already had most of its high quality nutrition removed by the animals that produced it. For the very best results from this food it should ideally be ground up and mixed with water. This simulates the chewing action of the animal and breaks the food into tiny pieces so the bacteria can eat it more easily and produce methane gas quicker.

All of these biodigesters do need to be situated in a sunny spot so that the heat from the sun warms up the bacteria enabling them to function and produce the methane gas for you. It is also a good idea if possible to place it close to the area that you are going to use for the cooking. This will cut down on the price of piping material to get the gas to the stove and it is also close to where the food scraps are usually produced.

You have been shown four types of urban biodigesters in this chapter that you can easily build yourself in your back garden. This should be enough design choices to get you started on building your own biodigester for any urban garden. Choose the design that you like best and get stuck in.

Important note: Any second hand containers that you use to make a biodigester with must be thoroughly washed and totally clean before you use it. Any oils or chemicals that were in the container before you bought it have the potential to kill all your bacteria as soon as you put them into the container. Ensure that it is a totally clean container and if possible go for containers that held fruit juice or vegetable oil that will not be toxic to your bacteria.

Chapter 5

Basic safety devices

Now that you have the basic designs you will need to put some thought into the basic safety requirements of your system. To this aim the following chapter contain the basic safety features that will be required on your biodigester. Always have these basic safety features built into your design of digester.

The water trap

The problem we have with taking a gas from a moist damp environment is the fact that a fair bit of moisture vapour can be carried along with the methane gas as it travels along the pipes. This water vapour can condense on the inside of the pipes and reform itself back into a liquid. A small amount of water won't affect it much but the problem is that this water accumulates at the lowest spot of the gas pipe forming a restriction or a total blockage of the gas.

Water vapour when it condenses will always fall down towards the lowest point in the system and pool there. Unfortunately if you get enough of this water pooling in the pipes then it can form either a restriction for the gas or a total blockage of it.

A sign that a water blockage is starting to occur is when you see the flame start to die down and flare up again as you are trying to cook with it. You could almost say that the flame starts to pulsate. This is caused by the gas which has been restricted by the water which starves the flame of sufficient gas, which can make the flame die down, and then as the gas manages to bubble its way past the blockage it gets pushed upwards towards the gas nozzle by the pressure of the water causing a sudden rush of gas that makes the flame bigger for a moment. If enough water forms in the pipe work then it can prevent the gas from flowing altogether.

To prevent this from happening the water can be removed from the pipes by the use of a simple water trap. This is an area for the water to drain into that can be drained of any excess water at a later date. There are a few different designs for these traps to choose from. This diagram highlights the basic elements needed for it to be effective.

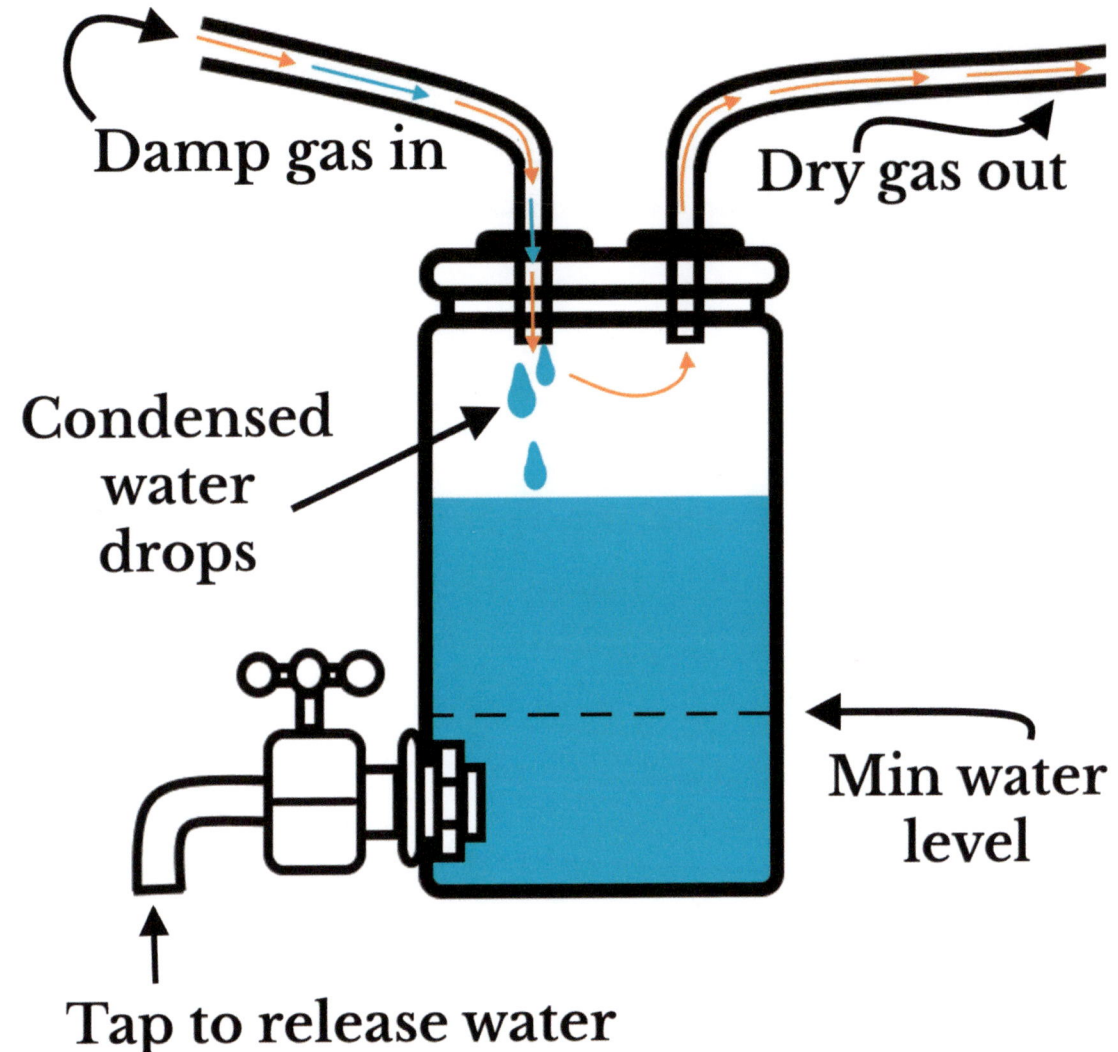

The tap is an easy type of water trap which is cheap to buy and simple to install. It does run the risk of letting air back into the pipes once all the water has been evacuated, which would be a very bad thing as air mixed with methane is explosive.

If you decide to use this type of trap then it's better for you to fix this tap to the lowest part of a clear plastic container so that you are able to view the level of the water inside the container. This way you will know when there has been a build up of water and you will know exactly how much you can release before there is a risk of letting air back into the container. If you always leave an inch of water inside the container (above the tap) then there is very little chance of any air being able to get into the container or the pipes ever. This is because the gas pushing down on the water is under pressure so the water should be forced out and no air should come in unless the water level reaches the tap exit point.

The last design of water trap was a simple two pipe design, one pipe in and one pipe out. Water can come in one pipe but gravity ensures that it cannot leave via the exit pipe. These water traps can also be designed using only one pipe as you will see in the following diagram which utilises only one pipe and a T-Junction.

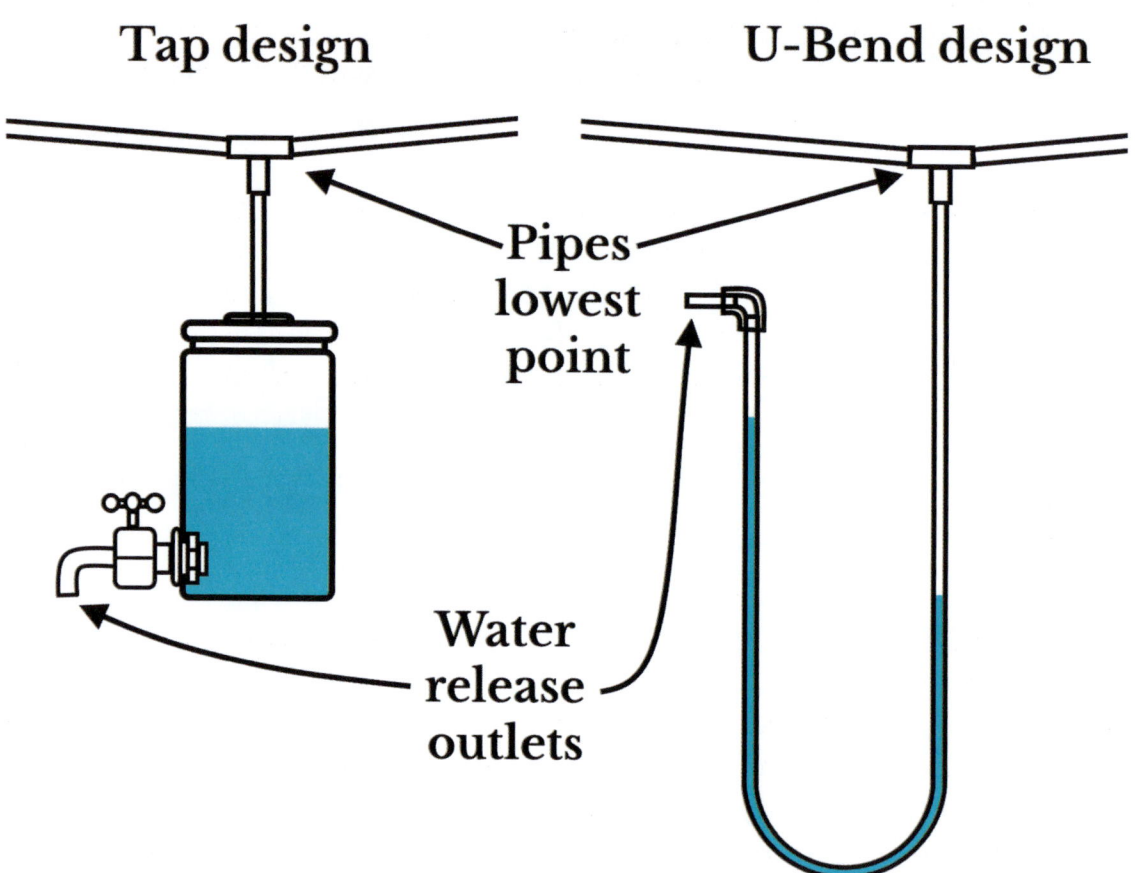

Because the water vapour that has condensed in the pipe runs down hill it should drop into the water trap and the dry gas should carry on uphill towards the stove etc. If using the tap sounds like hard work, especially if you have a lousy memory and may forget to drain it, then you may want to consider the U-Bend system.

The U-Bend design is an automatic water release system that allows any excess water that drips to the bottom of the pipe to be release from the over flow water release outlet. The pressure of the methane gas will raise the level of the water on the exiting side of the tube but so long as the water release outlet is higher than the level of the water on the methane side then it should equalise at its own level just below the outlet hole. Whenever more water drips down the inside of the pipe the water level by the exit pipe should raise up. If enough water enters the pipe then the water should drip from the release outlet automatically getting rid of any excess water. The height of the water release outlet may need to be raised higher depending on how much pressure is on the methane gas inside the storage section. But once this height has been set the pressure shouldn't really change much.

You do need to ensure that the water levels are checked regularly to ensure that the water hasn't evaporated etc. If the water isn't kept inside the tube all of your methane gas will escape into the atmosphere, so check it regularly. Thankfully this is the only thing that needs to be checked on this type of trap.

Flash back arrestor

A flash back arrestor (also called a flame arrestor) is a safety device designed to prevent the flame on the gas stove from going backwards inside the pipe itself and making its way to where all your methane gas is stored. The flame can actually travel along the inside of the gas pipe at twice the speed of sound so it can happen unbelievably quickly.

The DIY flashback arrestor is a very simple device to construct because all that it consists of is a section of pipe that is crammed full of bronze wool.

Most people are familiar with wire wool which is sold for scrubbing pans with, but this is not recommended for making a flashback arrestor because it can actually burn with the flame. If you place a lit match to a wire wool pot scrubber it will actually start to burn with the flame, so it is best not to use these.

The best wire wool to use will be bronze wool because this simply does not want to burn even if you put a flame directly beneath it. It is readily available to buy online from eBay, Amazon and most marine equipment suppliers. When it is crammed into a section of pipe all of the methane gas can pass through it without a problem, but if a flame ever tries to get passed it then it will be stopped in its tracks.

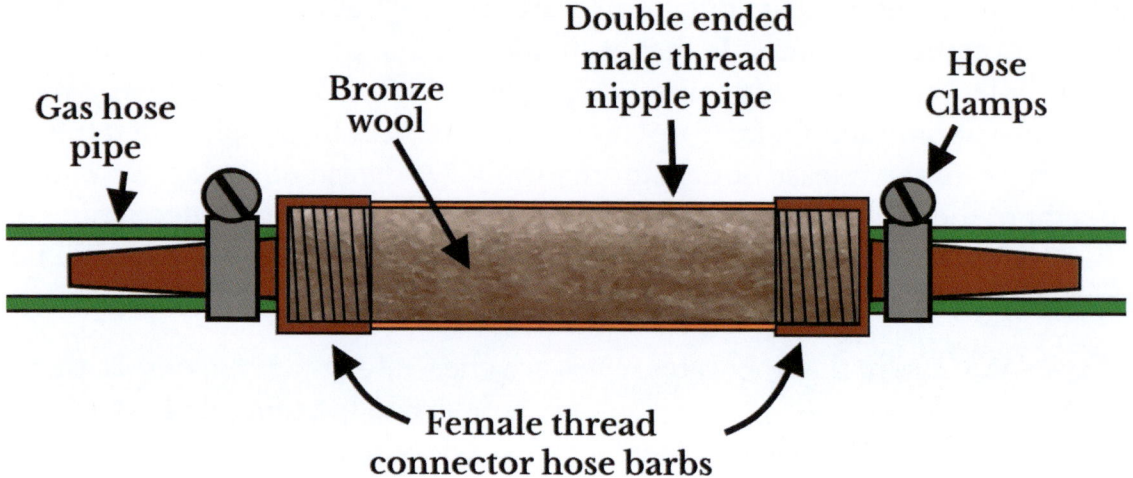

The bronze wool has a very large surface area that actually absorbs the heat from the flame which helps to cool it down and help suffocate the flame itself. The flame cannot progress fast enough through this maze of holes before all the heat and air is removed from the flame, this then extinguishes the fire. In normal conditions the bronze wool will not hinder the flow of gas through the pipe but a naked flame is a totally different matter and it should hopefully stop it dead in its tracks.

This should be positioned just before the gas pipe is attached to the stove. This way if there is a flame drawn back into the pipe it can only travel a very short distance before it is stopped.

Some people have had success using aquarium filters on their flashback arrestors but after hearing about a few of these crumbling and cracking over time I prefer to stick with the bronze wool.

Old style pebble flashback arrestor

Another older type of improvised flashback arrestor is the pebble trap which is very similar in function. The main difference is that it uses pebbles to trap the flame and extract the heat from it as it tries to pass through it. I have to admit that I have never tried using one of these but in theory it should function very well. I am not advising that you use it I am merely telling you of its existence.

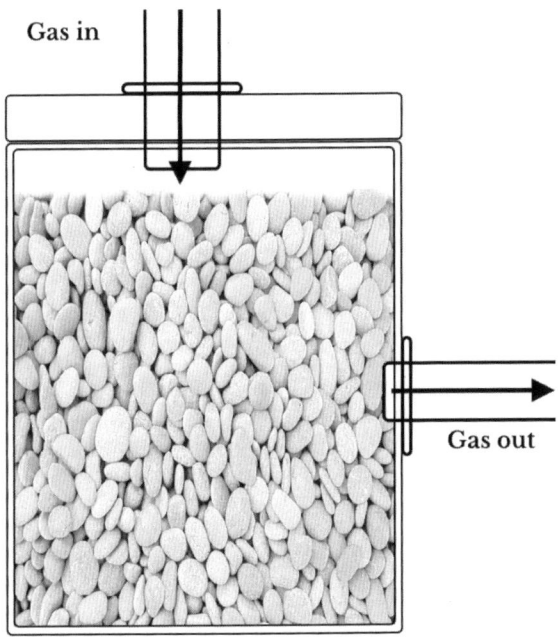

The gas can flow through the pebbles but a flame would have great difficulty navigating its way through the myriad of holes between the pebbles before losing all of its heat and oxygen first.

Important note: All joints must be made 100 percent airtight. Any joint that hasn't been sealed properly can allow gas out and air into the pipes, which would give a severe risk of explosions occurring. Every pipe, every container and every single part of the system must be sealed properly and checked for any leaks.

A cheap and simple test is to place a thick washing up liquid around the joint and look for any bubbles that form in the washing up liquid as the methane gas escapes.

Gas taps & valves

Controlling the flow of gas from one area to another is vital for your safety and everyone around you. A gas valve allows you to stop or restrict the flow of gas through the pipes and storage areas.

There are various different types and designs of gas control vales that you can buy. It is up to you to find the shape and style that will fit your design of biodigester. The diagram below shows how a typical gas shut off valve works.

This diagram of a see through valve shows how a large ball bearing inside the valve rotates to reveal a large hole drilled through it. If the hole is in line with the pipe the gas will flow through it. If the hole isn't in line with the pipe it will stop the flow of gas. Anywhere in between will allow as much or as little gas to flow through the pipe as you prefer. These are very cheap to buy and can be bought at most plumbing or hardware shops and online very easily.

Ideally these need to be placed on every part of your system i.e. one on the digester, one on the gas storage area and one before the cooking stove. This allows you to perform maintenance on any part of the system because you are able to close any of the gas taps ensuring that no gas will escape & that no air will get in.

Chapter 6

Storing biogas

It is all well and good being able to produce your own biogas but most digesters only have a finite amount of space for storing the methane gas inside the digester itself. You may be able to use it as needed but if you find that the amount of gas is starting to build up then you are going to have to find somewhere safe to store it.

Never be tempted to simply release excess methane into the atmosphere because it is a potent greenhouse gas which is about 25 times worse than CO2. This means that you only have two viable choices, you can either burn your excess gas or store it for use later. Storing your excess gas will always be the better option.

You have a few different choices where you can collect the gas that is produced by the biodigester. One choice is to use a large inner tube to collect the gas that comes from the barrel. The inner tube inflates as the gas is produced and stores it inside it till you are ready to use it. If the inner tube is full then the gas will be under pressure so cooking with it should be fine as is. If the inner tube isn't full then you can increase the pressure on the gas by placing a reasonably light weight on it to push the gas out.

Check that the inner tube that you have is totally airtight by filling it with air and holding it (or sections of it) under water and look for air bubbles leaking out. Ensure that you remove all of this air by tightly rolling the inner tube up before you start storing methane inside it.

If you intend storing the gas inside an inner tube but you want it so that you can disconnect it from the digester and take it elsewhere then you will need to leave the valve on the inner tube connector. This is so the methane doesn't escape while it is being transported. If you are going to have the inner tube permanently connected to the digester so that the gas can flow into the inner tube when it is produced, and out of it when it is being burnt, then you will need to remove the valve from the inner tube connector. This means the methane gas can move freely between the digester, the inner tube and the cooking device.

Low tech gas storage devices

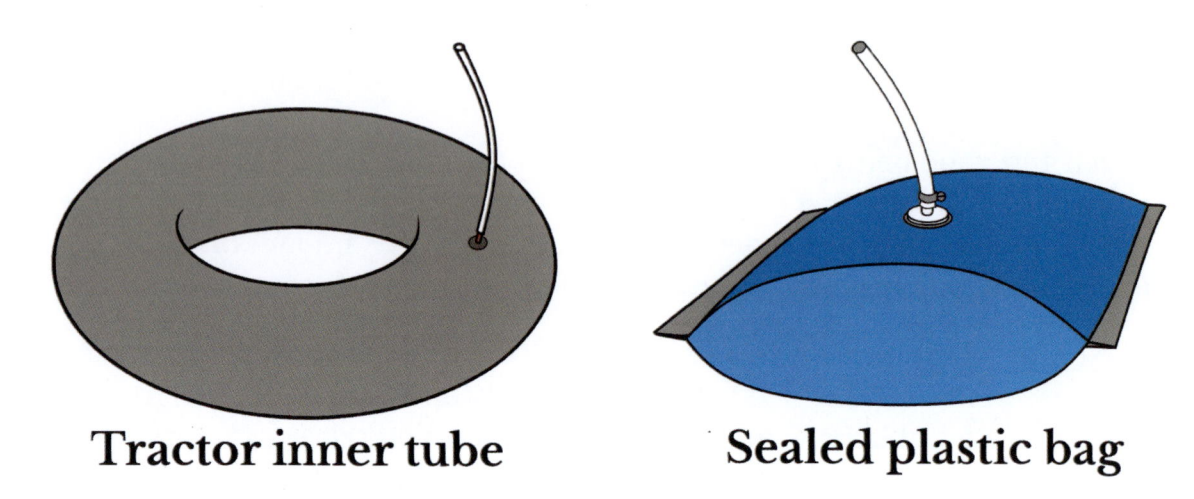

Tractor inner tube

Sealed plastic bag

Another choice which I haven't done personally but a lot of other people have is to store the gas inside large thick airtight plastic bags. They simply attach a tank fitting to the plastic bag so they can get the gas in and out of the bag and then they allow it to fill up from the digester. These bags would have to be very well sealed to ensure a gastight/airtight seal on it. People have sealed these bags using ductape (though I am not sure that I would risk this) in third world countries and used them to store and transport this gas when they need a mobile supply of gas or if they intend to sell it to other people. A very light weight (with no sharp edges) can be placed on top of these to increase gas pressure if needed. All joins and pipe attachment points must be 100% sealed to prevent gas leaking out or in.

Another option people have used for gas storage are weather balloons, because they can expand to quite a degree to store a lot of gas. They are quite durable and strong and they create their own gas pressure as they expand so no weights are needed.

The safest option for storing the gas is to use a floating drum style holder similar to the ARTI style biodigester. This is a fairly safe and robust collection device that tells you exactly how much gas is stored inside it by how high it rises out of the water. Again if you find that the pressure is a bit low all you need to do is place a small weight onto the floating barrel and this will compress the gas forcing it out. It consists of two barrels, one of which is slightly smaller than the other.

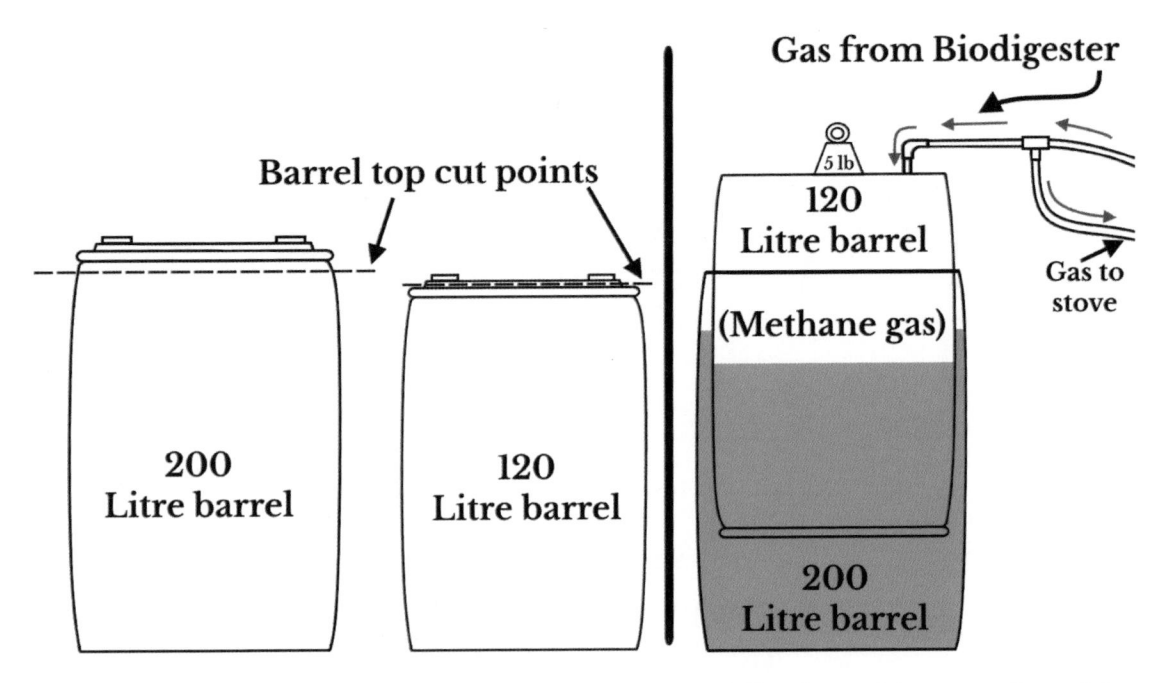

The tops are removed from both barrels and the smaller barrel is turned upside down and slid inside the bigger barrel. Because the bigger tank is filled with water and the gas is fed into the inside of the smaller upside down barrel the methane gas cannot escape and no air can get into it either.

This is one of the safest and easiest storage methods to manufacture yourself. Gas can be fed into the storage barrel and stored underneath it till needed. The bigger the barrels you can buy the more storage capacity that you will have inside it, but the more the barrels will cost.

You can build a frame around this to prevent the barrel from tipping over once it raises too high if it contains a lot of gas or you can simply place the inner barrel inside a much taller barrel because this will then act as the outer frame and look much better.

Using up spare methane gas

Having huge amounts of excess gas is very unlikely with a small biodigester but on the remote possibility that you have lots of spare methane then you can always use the excess methane to power a generator and either store your power in 12 volt batteries or simply utilise the electric for your daily needs. You will find that you have no problem utilising excess methane when you do this as the electric generator can quite happily munch its way through quite a lot of methane.

Generators convert liquid fuel into a gaseous state before they burn it so they are only too happy to run on the pure methane gas itself. This means that you not only have an adequate supply of gas for cooking or heating but now you have extra electricity as well

Before you start using the methane gas

When you first start your biodigester from scratch the first gases that you will get out of the digester will either be the remaining air or CO2. This is expected so don't be too disappointed at the fact that the gas that comes out of it for the first week or two simply won't burn, after these two weeks you should be fine.

Any air and CO2 gas in the system should be bled out as soon as it becomes pressurised because the very last thing you want mixed with your methane gas is air because it is just so dangerous. It is best to bubble the end of the gas pipe that is submerged inside a bucket of water this guarantees that no air can get back inside the digester itself, but the gas can escape when it starts to be put under pressure.

One thing that you must remember is the fact that air is the enemy here. Any pipes that you have connecting up to your storage devices can fill with air unless you have valves on them to stop air leaking into them each time they are connected or disconnected from a storage device. Seal it up and keep it sealed.

Chapter 7

Cleaning dirty gases

The gas that is produced by your biodigester will contain extra unwanted toxic gases. One of these is called hydrogen sulphide. You can usually tell when this gas is present because it smells like rotten eggs (or to be cruder about it the gas stinks of smelly farts).

Hydrogen sulphide is a very damaging gas to humans & pets because it can cause nausea, watery eyes, headaches, poor memory, damages the lungs, and a host of other medical complaints. The worst part about this gas is that it destroys your sense of smell after a little while, so you will be unable to tell if the gas is present meaning that you will have no warning that it is still around you. This is why you should deal with hydrogen sulphide gas immediately when you first smell that rotten egg smell

There is thankfully a very simple way to remove most of the hydrogen sulphide gas using a larger version of the flashback arrestor. The only difference is that this time we will be using steel wool (also called wire wool) because we need to use a metal that will rust. Steel wool will start to rust once it is wet creating a layer of iron oxide all over the surface of the metal. Iron oxide for those of you who don't know is commonly referred to as the common or garden "rust" that forms on the surface of steel and iron when it gets wet. It is the chemical reaction between the hydrogen sulphide and the iron oxide that neutralises the hydrogen sulphide and gets rid of the nasty smell.

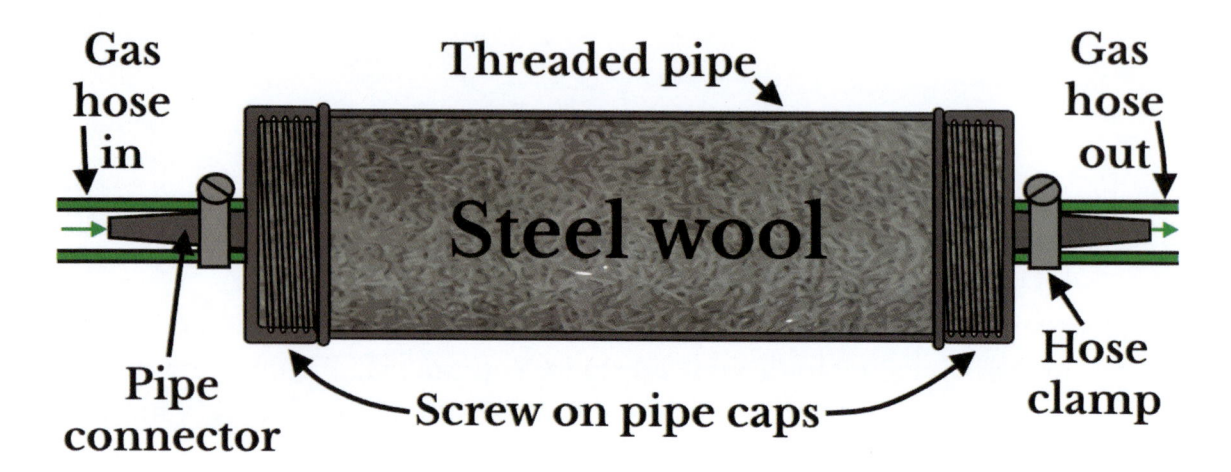

Gas hose in → **Threaded pipe** — **Steel wool** — **Gas hose out**

Pipe connector — **Screw on pipe caps** — **Hose clamp**

It is best to construct the filter out of a 4 inch PVC pipe with two screw on caps if possible for easy maintenance. Ideally it will be fixed into a vertical position so that the gas can come from underneath and will have to pass all the way up through at least one foot of wire wool (two feet is better) filled pipe before it can go any further. The hydrogen sulphide gas has to weave its way past the rust covered steel wool which has a huge surface area. The hydrogen sulphide should chemically react with the rust (iron oxide) neutralising the gas and getting rid of the rotten egg smell.

This wire wool will reduce in size over time so it is recommended that you check this every few months in case it needs renewing or topping up. This is why you are recommended to use (airtight) screw on caps on the pipe to allow you easy access to the wire wool when you need to change it over. If at any point you find that you can smell the rotten egg smell again then you have probably waited too long to change the wire wool and it should be replaced immediately.

Your hydrogen sulphide gas scrubber is best located directly after the methane gas exits from the biodigester if possible, this way the gas has to pass through the scrubber before it gets to the storage area or the stove.

Using a CO2 scrubber

The steel wool scrubber will hopefully take care of most (if not all) of the Hydrogen sulphide gas but it will do very little to cut down on the Carbon dioxide (CO_2) gas that is also in the biogas.

To help reduce the amount of carbon dioxide that you release into the atmosphere and your home we can build something called a bubbler. All that a CO2 bubbler does is pass the gas under the surface of the water and make it bubble its way through the water itself. As the gas is bubbled through the water some of the CO2 gas is absorbed into the water itself creating a very week acid called carbonic acid. Although it may not remove all of the CO2 it will remove a fair amount and it does do a better job than not bothering to filter it at all.

There should be enough pressure on this gas as it emerges from the digester to push it under the water. How far under the water it will push this gas depends on the amount of pressure that your digester is producing. It doesn't need to bubble the gas very far under the water itself in order to get the CO2 gas to be absorbed by the water. The deeper it can push it down into the water however the longer it will spend in the water and the greater the amount of CO2 is likely to be captured.

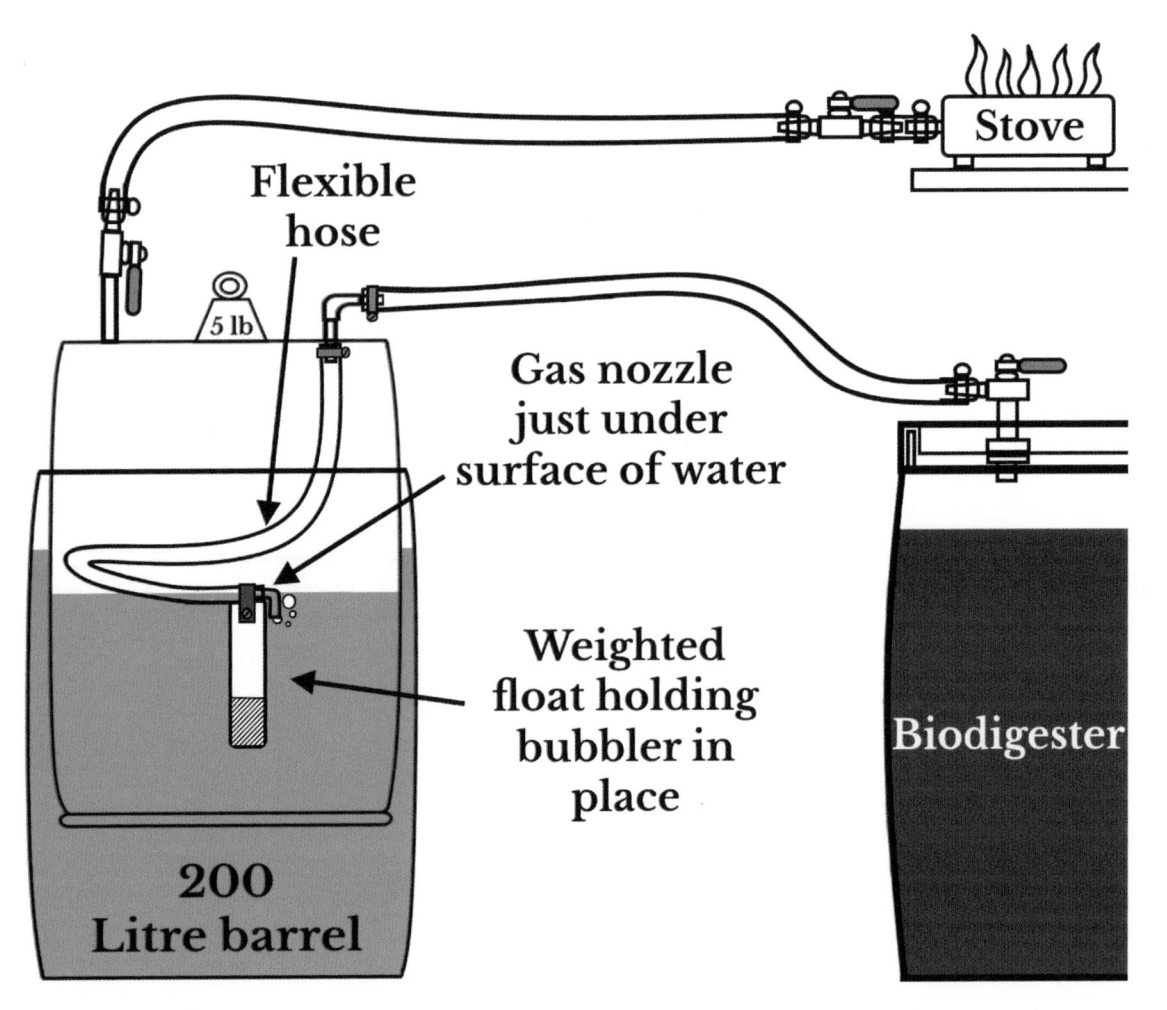

The gas is fed into the smaller upside down barrel through a flexible hose that brings the gas down to the surface of the water inside the barrel itself, where it is to be store for use later on.

The hose is attached to a weighted float that holds the end of the hose just under the water. This means that the gas has to bubble its way through the water before it can enter the inside of the barrel. When the height of the barrel changes the flexible hose ensures that the hose and the bubbler stay in exactly the same place in the water and that the methane has to continue bubbling through the water before it can enter the barrel.

Small note: Performing this bubbling process should also help to remove any hydrogen sulphide that may have escaped the previous filter.

Chapter 8

Bottling the gas

Bottling the biogas inside LPG canisters can be done but it is problematic and generally speaking not worth the effort or the expense for most people. I personally have not done this because I have no need to do it. All my needs for methane gas such as cooking and heating etc can be accomplished using a standard gas storing device such as an inner tube, a plastic bag or a floating barrel type storage setup. Because compressing the methane down to usable levels for use in vehicles etc can be very difficult to do, it is too much like hard work and expense to make it worth my while.

I did do a lot of research on this subject before I ruled it out. The information about this is included here for you. I am happy to share it with you, but it is research which I personally haven't tested myself, so it is given to you on an untested basis and you will have to check over the details of this information and verify its validity if you want to use any of it yourself.

Biogas gas will compress for storage inside an old LPG tank quite easily but it is the other gases that are with it that pose a bit of a problem. It is vital that the gas is scrubbed clean of these other gases that it contains before it is compressed. Whilst cleaning the gases for normal use is fairly straight forward with the filters you have already been shown, cleaning the gases for compressing it into metal LPG bottles however is a totally different ball game. The cleaning of the gases will have to be a lot more thorough if you don't want these unwanted gases to cause any problems.

You should already know that the main unwanted gases inside biogas are water vapour, hydrogen sulphide & carbon dioxide. And you have seen how to remove these to some degree for the basic use of the gas when it is stored using low tech methods (inner tubes etc). When you wish to compress the gas inside steel LPG bottles things change and these gases will have to have all of the impurities removed, not just enough to get by.

Removing the Hydrogen sulphide

We have seen how we can remove hydrogen sulphide gas by running the biogas through a wire wool scrubber to neutralise it. Removing the hydrogen sulphide

(H2S) gas is necessary because it will condense on the inside of the metal canisters and corrode the metal.

This part is relatively easy for you to accomplish. The basic steel wool filter should still work for this part of the cleaning process although having a second one of these to catch any remaining H2S that escaped the first filter may be a sensible option. Also the slower the gas moves through the filter the longer it will have to come into contact with the rusty metal giving a more thorough cleaning of the gas.

Moisture removal devices

Water vapour as you were shown earlier will be mixed with the biogas. You do not want water inside your gas tank at all, so removing the water vapour before it gets to the tank is a must.

Cooling the gas can help to make it condense into water droplets, then the water trap will function even better at separating the liquid from the gas. This is by far the easiest way, but some moisture will also be removed from the gas when it is pushed through the CO2 filter, which is covered next.

Removing the carbon dioxide

We have already seen how to remove some carbon dioxide gas from our biogas by bubbling the gas through the water so that the water would absorb some of the CO2 from it. This is okay for gas that we intend to burn straight from the storage device but there may still be too much CO2 left in the biogas for us to try and compress it down for use inside a gas bottle.

Carbon dioxide gas won't burn and also doesn't compress very well either so removing this is a must. Removing all of the CO2 can be an expensive problem but you can use cheaper methods and materials to get similar results.

Dirty gas in

Clean gas out

Biogas bubbles

Aeration stone bubbler

Milky coloured Calcium Carbonate formation

Calcium Hydroxide solution

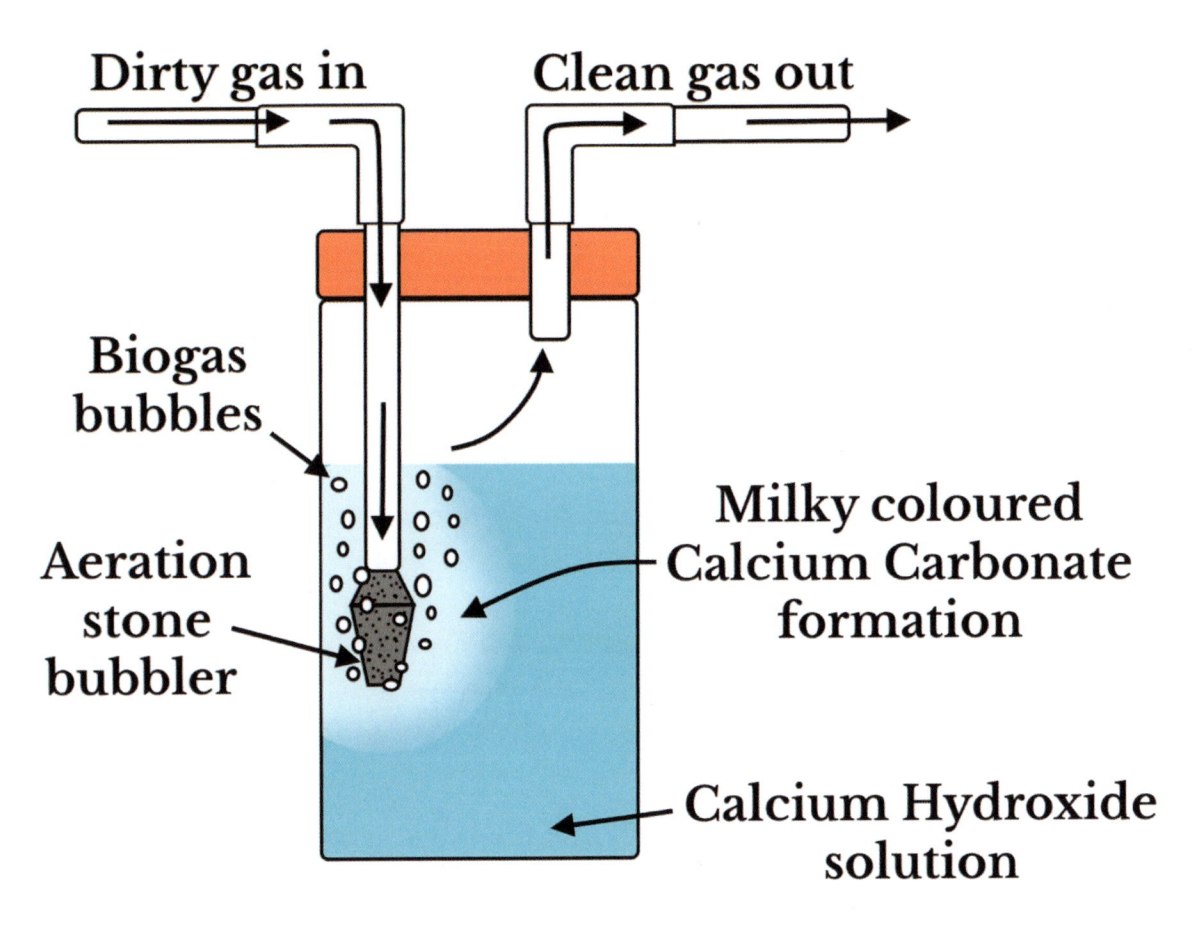

Using a combination of the bubbler & the water trap you are able to strip most if not all of this CO2 out of your biogas quite easily. In this combined bubbler water trap design (shown above) you use an easy to purchase chemical called Calcium oxide which will help the water lock on to and absorb the CO2 gas better. The CO2 gas is converted into calcium carbonate, which to you and me is better known as chalk. You will however have to remember to change these chemicals on a regular basis for it to be effective.

An aerating stone from an aquarium on the end of your bubbling hose can be very good at dispersing the gas into a myriad of smaller bubbles. The smaller the bubbles are the greater the surface area of the gas will come into contact with the water mixture itself, meaning that the liquid can capture more of the CO2 inside the water & chemical mix. The gas may need a slight amount of extra pressure behind it to pass through the smaller holes but the results are worth it.

Calcium Oxide

Calcium oxide (CaO) known by most people as quicklime is created by burning limestone until all of the CO_2 gas has been released from it, turning what is left into a fine white powder.

Calcium oxide is colourless and odourless in its dry form but when it is mixed with water there will be an exothermic reaction (meaning that it can get very hot) and poses a risk of burns etc so care must be taken when mixing or using it. When calcium oxide is mixed with water its chemical formula changes and it then becomes known as Calcium hydroxide with its chemical formula changing to $Ca(OH)_2$.

Warning: Be warned that this mixture gives off an exothermic reaction when mixed with water and will literally start to bubble and boil as soon as you mix it with water, real care must be taken to avoid burns etc. The dry powder reacts to water so if you have wet hands or some dust gets into your eyes then is can burn you pretty badly. So wearing gloves and safety glasses would be a sensible precaution. Keep the Calcium oxide in a dry sealed waterproof container so it cannot get wet until you want it to.

When you are ready to filter your gas you mix the calcium oxide with water which will give out a lot of heat at first but then it will settle into a nice clear liquid which looks just like water. This new watery mixture now called calcium hydroxide is commonly known as slaked lime. It can also be called slack lime, hydrated lime, caustic lime or builders lime along with a few other names.

Simply bubbling the biogas through this mixture should remove CO_2 gas so much better than just plain water would. Some moisture and any remnants of H_2S gas should also be removed when the biogas is bubbled through this mix so it can be a very effective filter medium to use.

Calcium oxide which then becomes your calcium hydroxide (when mixed with water) is a very cheap and readily available ingredient that you can buy in order to strip the CO_2 from the biogas for you.

When the CO_2 gas is passed through this solution it reacts with the calcium hydroxide and gets converted into calcium carbonate and the water solution turns a milky white colour

Sodium Hydroxide

Another chemical option you can consider is NaOH Sodium hydroxide, which is also known as lye or caustic soda. This chemical is dissolved in water and will quite happily absorb moisture and carbon dioxide from the biogas that is bubbled through this solution. This chemical & water mix can however cause severe chemical burns so extreme caution must be taken if you are considering using this.

Boosting its effectiveness

No matter what chemical you decide to use to filter the CO2 from the water you must ensure that this chemical mix is changed every time you clean the gas ready for bottling to ensure that the chemicals are working to their utmost capacity and are absorbing as much CO2 as possible.

Compressing the gas into LPG tanks

The force that the gas exerts on the inside of the bottle is measured in PSI pounds per square inch or sometimes PSIG (the G stands for gauge). This is an important measurement because the LPG tanks that you are using to store your biogas are rated at a certain level before it will rupture or explode. Generally speaking you should never get your gas pressure anywhere near this rating so this shouldn't be a great problem for you, but it is vital that you limit the amount of pressure that you put into the LPG tanks. If you are using old recycled gas tanks they may have weakened and corroded internally over time so they may no longer be strong enough to pressurise to anything like the rated PSI value any more. So choose decent bottles and look after them.

The temperature that the gas tanks are stored in can affect the internal pressure of the tank itself. This is because the gas steals heat from the walls of the metal tank in order to convert itself back into to a gaseous state. The hotter the tank is the greater the pressure inside the tank.

LPG gas is stored inside the tanks in a liquid form with a layer of gas vapour above it. When some gas is used the lower pressure allows the liquid to convert back into a gas to refill the available space above the liquefied gas. Liquefying biogas is not an option for you because you won't have the equipment or technical expertise to do this. You can however still compress the biogas inside the LPG tank up to a certain pressure so it can be used later on.

Some DIY biogas enthusiasts have been able to compress the biogas down to around 150 to 200 PSI pressure inside a metal gas canister using certain types of compressors. A lot of these DIY biogas people have had a lot of success with compressors from refrigerators. A standard air compressor is not usually recommended though because it was never design to operate with explosive gases.

It is very important to purge the pipes of air before you start to put methane into the tank because you never want a mix of air & methane going into the tank. To do this you disconnect the pipe that is attached to the gas tank and pump the compressed methane through this pipe. Once you are sure there is no air in the pipe you can close the valve and reconnect it to the gas tank ready for it to be filled up with methane.

The maximum amount of pressure you should ever go to using an LPG tank is a maximum of 200 PSI. The pressure limit for an LPG tank is usually set at a maximum pressure of around 365 PSI before any pressure release valve opens (if the tank has one). The absolute maximum pressure rating is around 1000 PSI before it explodes (this can vary for each different tank and the ambient temperature that the tank is stored at), but as stated earlier you should never go anywhere near the maximum level .

A reasonable amount of pressure to aim for is around 160 PSI inside your tank as this will give you a reasonable amount of storage inside your tank for general cooking heating and generator requirements. A PSI pressure of 160 is quite easily achieved using a cheap or even free refrigerator compressor unit.

The refrigerator compressor seems to be a favourite amongst the DIY biogas makers because they are so simple to get hold of and they are so cheap. They are not overly complicated to use and they are totally sealed units so no air can get in and no methane can get out. It is only when you have a mix of air & methane that there is a chance of an explosion. So long as you purge the system of any air then the risk of any explosions occurring should be very, very small indeed.

Using a compressor to squash down explosive gases does come with some risks so it is up to you whether you want to take these risks. I personally didn't find the need to compress biogas in this way and most other people won't need to either.

Warnings: All devices & tanks must be fitted with pressure valves to prevent the compressor from over filling the tanks which would make the tank rupture or explode.

As I stated earlier this information is from untested research I found in the early stages of my biogas training. If you decide to use any of this information I urge you to double check that this information is valid & correct before you use it. If you do use this information then it is done so at your own risk as I have not verified the technical aspects of compressing biogas, because it is not something I ever intend doing.

I personally would advise & urge you to simply store your biogas in the normal low tech way especially in the beginning of your biogas learning phase. Storing gas in the low tech way is much safer for you and your family and entails much less risk.

Chapter 9

Keeping them warm

The bacteria are very similar to you in a lot of ways. They like to eat the same kind of food that you do (and a few foods that you don't) and they generally like to live at the same temperatures that you enjoy. In the summer they are toasty and warm and they will happily munch through their food and produce plenty of gas for you. In the winter however they get cold, they can become dormant and could possibly even die if they are not given some degree of protection.

It is vital that the biodigester is insulated in the colder months to help keep the bacteria alive and producing gas for you. Insulating the digester is an absolute must over the winter time. You would not leave the house in winter without a coat so why would you leave the bacteria outside all the time without a coat for them also.

Keeping yourself warm with insulation & heating inside your house is a must in winter so giving the bacteria the same degree of thought is also a good idea. Not that you would be plugging an electric heater into the back garden for them as this would just be dumb. There are however a few other options open to you if you want to keep the bacteria warm and help them to survive and thrive through the winter.

Insulation

Insulation through the winter is an essential for your biodigester to survive & thrive. If it gets cold then they will go dormant. So wrapping the digester with insulation is vital if you wish to keep a good supply of methane gas flowing and to help the bacteria to get through the winter happy & healthy.

You will find that most types of insulation can be used so long as the digester is weatherproofed properly. Rockwool or even fibreglass insulation wrapped around the outside of the tank will work fine so long as it does not get wet. As soon as water gets into the insulation it will lose its insulating properties and can actually conduct heat away from the digester itself, so this type of insulation must be kept dry.

Solid or rigid closed cell foam insulation can also be used and has the added benefit of being fairly weatherproof because it won't accept water inside it. This can also be bought second-hand from old house renovations so it can be fairly cheap to buy. It can be a very easy job to wrap an IBC tank digester with this stuff as they are square in shape, but it can be a bit more awkward using it on a round tank.

Spray foam is a relatively cheap and easy way of encasing the tank in a skin tight, gap free jacket that will keep the tank quite cosy for the bacteria and is also fairly resistant to rain and water ingress. The foam would have to be the high density closed cell type of foam to be considered totally water resistant. If you can house the digester inside a small shed or sheltered area then waterproofing isn't an issue and it will keep the digester much warmer throughout the winter.

Top & bottom

Don't forget that heat can be lost at the bottom and the top of the tank. So you will need to ideally raise the biodigester off the floor and onto a pallet, and then push insulation into the inside section of the pallet to keep the heat from escaping underneath the digester itself. A good thick layer over the top will finish it off and hopefully keep your bacteria very happy and productive.

Warming them up

You may also need to actually heat the tank up in really cold climates and there are a few ways of doing this depending on your technical skills and the amount of effort & cash that you wish to put into this project.

The electric heating element

Just like the immersion heating element that goes inside a hot water tank, you are able to buy heating elements that you can plug into your biodigester. These can be mains voltage which you would need to pay an electrician to wire up or you can buy a 12 volt version which you can instal yourself quite legally and safely. This means that you can power these in an off grid setup using only solar panels to power it. Wiring these up is fairly straight forward but if you don't understand electrics then you can simply pay someone to do it for you.

12 Volt heating element

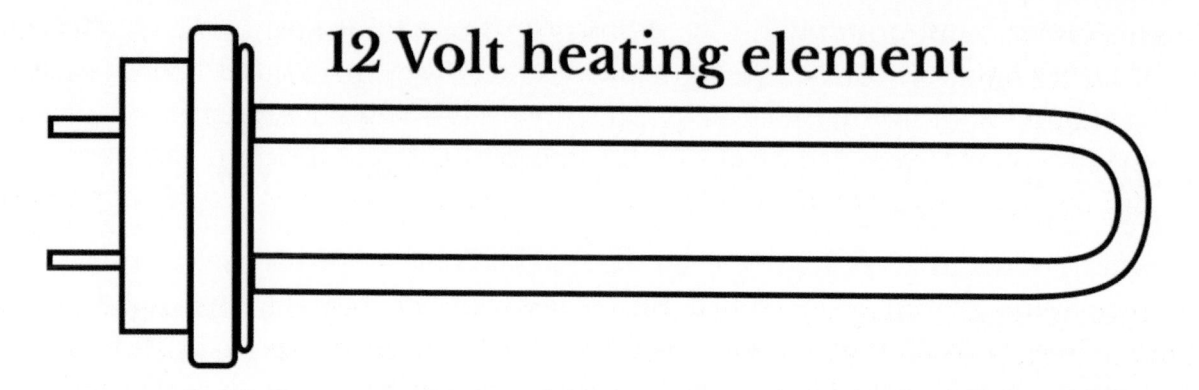

You will need a temperature control device if it doesn't have one built into it already to ensure that you don't cook the bacteria but these are easy to get hold of and install.

The low tech method

There is also a very simple low tech method of introducing heat into the digester that only requires some very simple DIY plumbing skills. This method involves using a garden hose pipe which has been coiled up at the bottom of the digester. All that you have to do is connect the hose up to a warm tap and run a little hot water through the pipe. This will transfer heat into the digester raising the temperature level inside the digester itself. The insulation should help to keep this heat inside the digester, which will keep the bacteria happy. If you don't like to waste too much water you can simply pour a bucket of hot water into one end of the coiled hose pipe and allow the water to sit in the coiled pipe till all the heat has been extracted from it, vastly reducing the amount of hot water that you use each time.

Any warm water that comes from the other end of the hose once the heat has been extracted can be used on the garden or mixed with the food that you are going to feed the biodigester with, this way they get nice lukewarm water with their food rather than freezing cold water that would cool them down again

Insulation

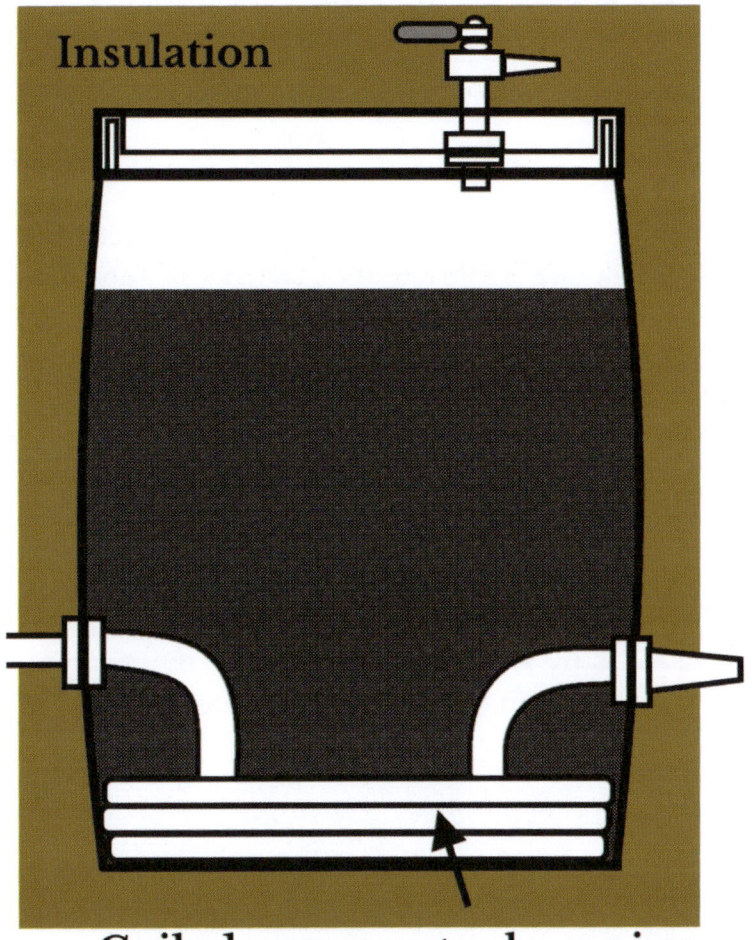

Coiled warm water hosepipe

Most people are capable of creating this heating system and if you are feeling creative you could even hook this up to a solar water heater which would constantly pump heat into the system in the winter months.

Another ultra simple solar heater is to simply install an old double glazed window on one side of the tank (the side that catches most of the sun) while the rest of the tank is insulated. The exposed section of the tank behind the glass is painted black so that it absorbs as much of the sun's rays as possible, thus heating up the inside of the tank. Some heat will be lost through the window when the sun isn't shining but generally speaking the solar gain out ways this amount of lost heat.

Chapter 10

Feeding the digester

Animal manure is a waste product that can be a problem to get rid of if you have an excess of it. Processing this waste into methane gas has the added benefit of processing the manure and converting it into a usable liquid fertiliser product that can be easily used on plants and vegetables. So if this waste product can be used and processed into something useful it would be silly not to do this. If you live in a rural setting and have access to this type of animal waste then it only makes sense to use it.

The urban design biodigesters are designed for use in cities and are fed high quality waste that has not been eaten by any animals as a daily food source. This high calorie uneaten food is packed with energy for the bacteria to munch on.

Food high in sugar, protein and starch will produce much higher quantities of methane gas than a manure powered digester would. The problem with using manure to generate gas is that you need an awful lot of manure to produce a small amount of gas because it has already had most of its calories and nutrients removed by the animal. This means that in order to get a viable amount of gas production you need a very large digester. Although manure does produce gas & fertiliser it is actually a very low energy food for the bacteria. This is because most of the goodness that is in the food that the animals eat is taken by the animal itself to grow and get strong. What is left over are only the waste products that the animal could not process. This means that most of the energy has already been extracted before your bacteria can get to it. They will happily munch their way through the energy that is still left inside it but they are only able to extract a certain amount from this left over material.

With using food that hasn't already been digested by an animal the bacteria can have a feeding frenzy and produce much more gas with a much smaller digester and a much smaller amount of food.

Manure also takes longer for the bacteria to process than food waste does. It may take up to 40 days for manure to be fully converted and broken down whereas the food scraps can be broken down within a few days.

Feeding the babies good food

You must think of the bacteria as children that must be cared for each and every day. They must be given fresh food & water daily if you want them to stay healthy and produce plenty of gas. As tempting as it may be to simply dump your scraps of food straight into the biodigester some thought has to be put into the food preparation first. To start off with you need to think about the types of food that you are going to feed the digester. Generally speaking if you are capable of eating this type of food then so is the biodigester. Most types of food are acceptable to a digester. They love bread, meat, fatty foods, cakes, fruit, vegetables and lots more. And the great thing is it doesn't have to be fresh food that is still edible. It can be rotten fruit, mouldy cheese, smelly stinky meat, stale crackers and any old disgusting food that can be salvaged from shops and restaurants that are no longer fit for human consumption.

Bad food: If there is a food that you yourself cannot eat such as bones, egg shells, peanut shells etc then it is unlikely that the bacteria will be able to eat these either. Although bones are organic in nature they are simply too tough for the bacteria to break down, as are the peanut shells etc. Any food that is too tough for humans will mean that it is generally not suitable for a biodigester itself and they should not be fed to the bacteria. If you do feed these types of food to the bacteria then they will not be able to digest them and they will simply sit at the bottom of the digester in a pile. This pile will simply get larger and larger inside the digester and eventually prevent new proper food being added to it. This will mean that you will have to empty and clean out the tank and start from fresh before you can start producing gas again.

You may think that garden waste would be inedible for the bacteria also but this isn't the case. The bacteria can eat garden waste such as grass clippings & leaves that humans would not be able to eat. Obviously the tougher the plant waste that you put into the digester the less likely they will be able to eat it, so use a bit of common sense.

Don't go dumping masses of any of these types of food at any one time. Just like humans the bacteria prefer a normal sized meal with a mix of different types of food that should be given to them at regular intervals.

Preparing the food

Once you have found the food for your bacteria to digest it will need preparing for them. Cutting, mashing or grinding the food up into small particles means that the bacteria will be able to digest it much faster. You are in effect chewing the food up for the bacteria like the animal would before the food gets down to the digestion area.

The larger the pieces of food are that you put into the digester the longer it will take the bacteria to break it down and digest the food. This would mean that the bacteria will produce methane gas much slower than they would if the food is broken down into tiny pieces. Smaller food means faster methane gas production so it is worthwhile doing this.

Breaking the food into smaller pieces

The cheapest and most tedious way is to cut the food into very small pieces which can be very tedious but it is free to do and is available to everyone who owns a sharp knife. The small chunks that are hand cut are not the perfect size but they will work fine so long as the pieces are cut very small.

You can spend a small amount of cash and buy a simple hand crank off grid food grinder and mince all the food before it is fed to the bacteria. This still involves a bit of work but it is better than chopping it into tiny pieces by hand.

By far the easiest but most expensive way is to buy a garbage disposal unit (also called an insinkerator) and simply dump any food scraps that you have into this. The food will be obliterated into thousands of tiny little fragments that the bacteria will adore. This machine will create the perfect mix of water and food for the bacteria to consume. You will need an electric connection to power it but it does make life so much easier for you.

Feeding it once in the morning and once in the evening is the ideal way to look after the bacteria and produce the maximum amount of gas.

The amounts that you feed the digester itself will depend on the size of the digester. A large scale digester will take many kilograms of animal faeces and water each day whilst your smaller urban digester will only need around 2 kg of food scraps (this amount is based on a 1000 litre IBC tank your digester may need less) mixed with water. Splitting the feeding times into a morning and evening feed will be much better and should produce gas fast and consistently throughout the day. If you can only feed it once per day then this will be acceptable, but twice a day is much better.

Each time you feed the bacteria they will start to digest this food and within about 6 hours they will start to produce gas from this waste. This 2kg of scraps should generate enough gas to last for about an hour of cooking time depending on the size of your digester and the health of your bacteria

Of course each time you add new food & water to the digester the more liquid fertiliser will come out of the other side of the digester itself. This is some of the best free high quality fertiliser that you can get your hands on and can be put straight onto the vegetable patch, so make sure you catch this and use it.

Supercharging the food

Although it is recommended to feed the bacteria a healthy rounded diet of mixed foods, the bacteria (like humans) love sugary foods and will respond to it with a lot of methane production. Sugar rich foods are extremely rich in calorific content so the bacteria will have a feast if they are given old cakes, fruit, fruit juice, spaghetti sauce etc. Anything that is high in sugar turbo boosts your production of methane gas. And like with humans the saying "a little bit of what you like will do you good" is quite right. Don't over feed the bacteria foods like this as this may cause other problems but if you get a supply of sugary food types don't be afraid to treat the bacteria occasionally.

Recharging the bacteria

The bacteria will feed & multiply very well inside the biodigester. But after many generations the bacteria can become tired and weak. So it can be very beneficial to boost the population of bacteria and introduce some new blood so to speak. Adding a fresh influx of new bacteria into the biodigester by adding fresh manure from time to time can reinvigorate the population of bacteria inside the digester itself. It helps to keep the entire system healthy and functioning at peak performance. A fresh dollop or two of manure every two to three months can see great benefits if you can manage to get your hands on some. The bacteria will digest the manure and the new bacteria will strengthen the population inside the digester itself.

Chapter 11

Checking the PH levels

If we utilise a basic scale used commonly in chemistry we can find the PH levels in your digester. This scale determines how acidic or alkaline the water is inside your biodigester. The scale runs between **1** and **14** with **7** being classed as the middle or the neutral point on this scale. Your biodigester ideally likes to be at a PH level of between **6.8 & 7.2** on this scale to be considered healthy. As the number on this scale drops below 7 the more acidic the water becomes with 1 being the most acidic it can be on this scale.

A lot of people overlook the checking of the PH levels until their digester suddenly stops working and the production of gas drops. It's easy to check the PH levels of the water that is inside the digester itself. You can do this easily by checking the PH level of the liquid fertiliser that comes out of the digester when you add new food & liquid. You should ideally check your PH levels at least once a week so that you can spot any problems before it becomes too late to do something about it.

Generally speaking you shouldn't have many problems with your PH levels if you feed your digester everyday because the water that you are using from the tap is at a good PH value already. It's always better to allow the water to sit in an open bucket for an hour or two, as this gives any chemicals that your water company may have added to it (in order to keep it fresh) a chance to evaporate from it, but this isn't essential.

This fresh water with a good PH value should help dilute any PH problems that may be building up inside the digester. This fresh water displaces some of the existing liquid every day and should hopefully keep your system healthy. If you forget to feed the bacteria for a few days then the water sits inside the digester and the acidity in the water can change to toxic levels for the bacteria. So feeding the bacteria daily is always the best solution. So remember to monitor the PH levels of the water inside the digester on a regular basis so you can spot any possible problems and fix them.

One major warning sign that your PH levels are out of balance in your biodigester is the fact that your production of usable methane gas will drop severely if it doesn't stop altogether.

You can check the PH levels by using specially treated strips of card that you dip into the water. These pieces of card change colour to indicate the PH levels. You can also buy a digital PH tester that displays the PH level digitally for you which most people would prefer. Which one you prefer is up to you. Both of these types of PH testers are available online or in most aquarium shops.

Increasing the PH levels

Ideally you can fix most PH level problems by adding daily amounts of fresh water but if this doesn't work then you can used a very common chemical that can be bought easily online. You can increase the PH levels of your digester quite easily by adding some "Soda Ash" also known as "Sodium bicarbonate" to your digester. This is a white odourless powder that readily absorbs water from the air so it should be stored in a dry airtight container.

Add a small amount of the soda ash to your food & water mix when you feed the digester and see how much it increases the PH of the water inside the digester. This will give you an idea of how much more you need to add over the next day or two to alter the PH levels back to where they should be. Don't be tempted to add multiple cups straight away as this may be too much for the system to handle and the PH level may spike.

Once you get to the correct PH level you can stop adding the soda ash and control the PH by regularly adding the fresh water mix each day. You will also notice that when the PH level is back where it should be that your methane gas production should increase to proper levels again.

Small warning: Soda ash get very hot when it gets wet so don't allow your bag of this to be left out in the rain.

It can also cause skin irritation, blisters & burns especially if your skin is wet when you handle it, so wear gloves. Your eyes are moist at all times so they can also get burns if this dust blows into them so wearing safety glasses may be an idea.

Read the warning labels and follow any health advice that comes with these products and consult with a medical expert if you have any incidents with these.

Lowering the PH level

If you find that your PH levels are too high then you can use an alternative chemical called Muriatic acid to lower the PH value of the biodigester. When the levels are too high a very small amount can be added to bring down the levels.

It is very important to only use a very small amount at a time as it packs a punch. A very small amount can have a big effect so use it sparingly.

It is also a very corrosive acid (hydrochloric acid) and as such great care must be taken when using it. The strength of different brands can vary greatly so check the concentration of the bottle.

This can be bought quite easily online also.

A quick fix

One very quick fix that may work for you if you don't want to add chemicals into the equation is to stop feeding them any solid food for a few days and simply use the liquid fertiliser that comes out of the outlet pipe on the digester to feed the digester with. There is very little food left in this liquid but there will be hundreds if not thousands of bacteria that got washed out with the liquid itself.

Sometimes doing this every feeding time for two or three days may inhibit acid production and help to fix PH issues, allowing you to resume feeding them again in a few days time once the PH levels have resumed to normal. This doesn't always work depending on what the problem is but it is worth a shot. If this doesn't work then you may have to result to using the chemicals.

Chapter 12

Converting gas stoves to use biogas

Warning: Modifying any gas appliance carries risks and may even be illegal in some countries. The work that is needed to be done to allow cooking devices etc to work using biogas is not a difficult thing to do by any means and virtually anybody would be able to do this without a problem. You are however altering a gas fuelled cooking device to do something it was not designed to do and it is up to you to ensure that the finished cooking device is safe and legal to use in your particular country. Officially in the UK any work carried out on a gas appliance needs to be done by a Corgie registered gas engineer although many, many people regularly change the gas jet (restrictor bolt) by themselves and don't mention it to any official body. The author of this book is not advising you to do this work yourself and does not accept any liability for any actions or consequences as a result of you altering your own (or other peoples) cooking devices in such a manner.

There seems to be far less restrictions when it comes to using camping gas stoves & LPG stoves so this is the type of stove that we shall cover in this chapter, but the instructions for camping stoves are basically the same for domestic stoves.

The pressure is on

The gas that gets supplied to you from your utility company flows through the pipes at a constant set pressure. The amount of gas that your gas cooker allows through to the part where the flame will be lit will be set for household pressure gas. Your cooking stove will have a restrictor to limit the amount of gas that is allowed through to the combustion area converting higher pressure gas into a lower pressure gas.

The pressure of the gas will be different for a household stove than it will be for a camping or LPG stove. The gas for a camping stove comes from a gas cylinder which means that it is under quite a bit of pressure.

Gas burner ring diagram

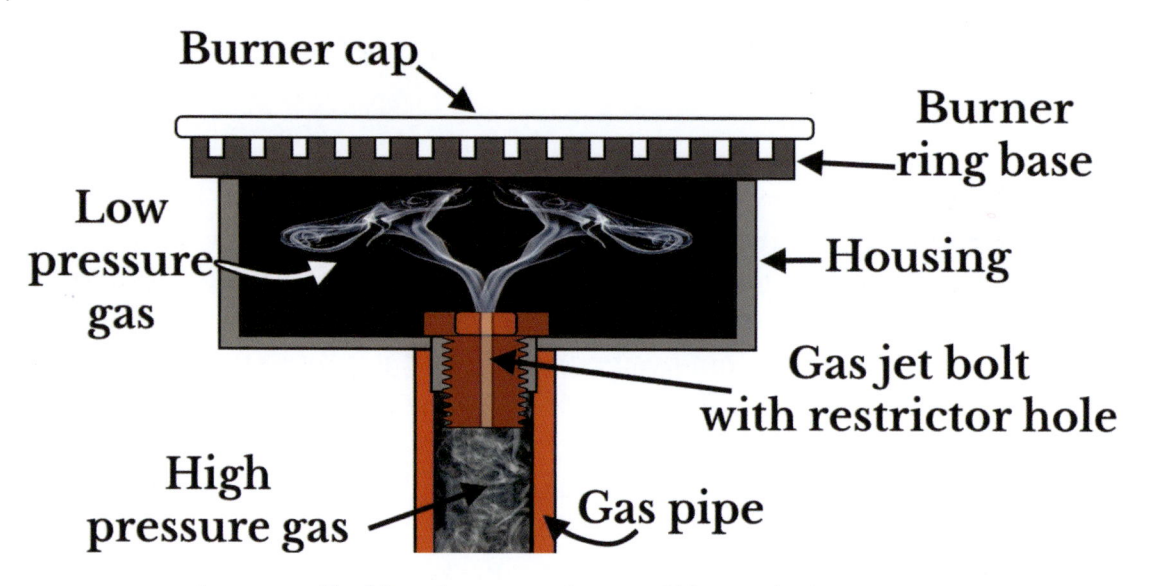

Burner cap
Burner ring base
Low pressure gas
Housing
Gas jet bolt with restrictor hole
High pressure gas
Gas pipe

The gas pressure is controlled by the use of a small brass bolt with a tiny hole drilled through it (a restrictor hole), called the gas jet. This means that the hole in a gas jet restrictor on a camping style stove will be a different width than the hole that is drilled into a standard household type gas jet restrictor.

You can actually buy a set of these gas jet nozzles to convert your standard household stove into an LPG stove so that you can use it off grid. The only thing different about the standard gas nozzles & the LPG gas nozzles is the diameter of the hole that has been drilled through the gas jet nozzles.

The hole through the bolt is called a restrictor because the width of the hole drilled through the bolt restricts the flow of gas through it. The smaller the width of the hole drilled through this bolt, the smaller the amount of gas that will be able to squeeze through the it at any one time. The higher the pressure of the gas the smaller the width of the hole has to be. And the lower the gas pressure is, the wider the hole through the bolt needs to be.

The gas from a biodigester has a very low pressure gas supply (a very small weight on top of your gas holder). This means that the gas would struggle to get through any kind of restriction placed in the gas pipe. This is why in most cases you can get away with removing this gas jet bolt altogether leaving quite a large hole for the low pressure gas to come through. Simply unscrewing the gas jet will usually solve all your problems with using the low pressured biogas that is coming from your low tech storage devices (inner tube, plastic bags or gas tank).

First ensure that your cooking device is unplugged from the wall (if it has a plug) and ensure that all gas supplies are shut off properly. You can then simply lift off the burner cap and burner ring base so that you can access the gas jet bolt. Then with a suitably sized socket (usually around 7mm) you can simply unbolt the gas jet bolt from the centre of the housing. Then you can simply reassemble the burner ring base, burner cap and then connect the biogas to the stove so that you can test how it works. Because there is no restrictions in place the biogas can flow at a good rate just like a normal stove would. You can control the speed of the flame by using the control knob on the cooker as you normally would whilst cooking.

Very important warning: Do not ever use this modified stove again with pressurised gas unless you replace the gas jets first as it will be very dangerous to use it without the gas jet restrictors in place (think blow torch right in front of your face dangerous). Without a gas jet restrictor in place the gas would race out at full pressure increasing the risk of explosions or severe burns.

Keep the original gas jets that you removed in a safe place in case you ever need to use this cooking device in the future with the gas supply that it was intended to be used with.

Chapter 13

Dangers & precautions

Because you are dealing with waste products such as manure and rotten food etc hygiene is a must. Manure contains viruses, bacteria & may also contain parasites. Whenever you have contact with the contents of the digester it is always a good idea to thoroughly wash your hands afterwards, particularly before you eat and drink anything. Generally speaking the fertiliser produced is a lot cleaner coming out than when it went in but even so it is always better to use good hygiene practices.

The digester produces three main gases Methane (CH_4), Carbon Dioxide (CO_2) and Hydrogen Sulphide (H_2S). Each of these gases comes with its own dangers and health considerations.

Methane: Methane is lighter than air so it floats upwards which isn't to much of a problem if your digester is kept outside. If it is inside a building then any escaping methane can float up to the ceiling and collect their creating the risk of an explosion if it comes into contact with a naked flame or a spark.

Carbon Dioxide: Again if the digester is outside then CO_2 is not that big a problem but if it is in a building then it can be an issue. This gas is the exact opposite of methane because it doesn't float upwards, it actually sinks down to floor level and literally sits there building up and getting higher and higher off the ground as it does. Higher than normal levels of CO_2 can cause headaches, dizziness, confusion and loss of consciousness. This is especially true for your pets that are at a lower level than you.

Hydrogen Sulphide: You can usually detect the presence of low levels of H_2S by the smell of rotten eggs. The problem is that with higher levels or after having contact with the gas for a little while it damages your ability to smell it, so you can be unaware of it building up into higher concentrations. It is a very poisonous gas that is corrosive and flammable. Humans do produce a small amount of this gas in their waste products which can be smelt when they evacuate their bowels. It is the smell that lingers in the toilet after your other half has been in there. This gas is also heavier than air so it will linger around the floor level until a door is opened and it is released into its surroundings. Although there is generally speaking only 1% hydrogen sulphide in biogas it is the smelliest portion of the gas so it is easily detected.

This gas can cause headaches, nausea, tremors, convulsions, damage the linings off the lungs, damage your ability to smell and lots of other problems. It was at one point used as a chemical weapon during World War one so if you detect this gas then ensure that it is dealt with straight away.

Generally speaking the dangers of poisoning from gases are minimal for outside back garden digesters, it is more of a worry for farms with huge pit digesters where people can quite easily pass out with the high concentrations of these gases. The dangers of fire or explosions however is present on all digesters so good safety practices should always be adhered to.

Chapter 14

Finding the prime suspect

Sometimes for no apparent reason your bacteria may stop producing gas and may even die. You may think that there is no reason why this should happen but there is always a reason why something has gone wrong with the digester itself. A healthy population of bacteria that have been fed properly should always thrive. If for some reason the bacteria have died then you should become the master detective and suspect murder! This may sound a little dramatic but it is important to find out why hundreds of thousands, if not millions of bacteria all died at the same time in suspicious circumstances.

You do this by first finding the murder weapon and then gathering together the most likely suspects and investigate where they were and if they have an alibi.

Ask yourself:

Has it been poisoned?

Is it possible that someone has put bleach or some other noxious cleaning fluid accidently (or on purpose) into the water that you have mixed with food for the bacteria. Maybe someone had cleaned the buckets out because they smelt and not rinsed the buckets properly afterwards. Maybe one of your children has fed the digester something totally unsuitable not knowing that it could harm it. There are quite a few possibilities on the poison front so it is down to you to try and discover if this is the case and who could be responsible (you may find out that it was you).

Has a silent killer crept into the digester?

One of the biggest possible killers for your bacteria is oxygen leaking into your container somewhere killing the bacteria. Oxygen breathing microbes produce CO_2 gas as a waste product, so if gas is being produced but it is not burnable then you should suspect murder by oxygen poisoning. If you get nothing but un-burnable gas then this is evidence that oxygen is your killer and it is down to you to work out how it is getting into the digester to commit the heinous acts of murder on the bacteria and fix the problem.

Killing the bacteria with kindness!

We talked earlier about not overfeeding or underfeeding the bacteria to keep them healthy just like you would do with a human. It is possible that you could be the killer in this detective story by giving too much or too little food, which can affect the PH levels of the water. Checking the PH levels regularly can ward off your overindulging or under indulging ways before it becomes a problem.

By feeding the bacteria no more than twice a day (morning & evening) and checking the PH levels regularly you can ensure that your bacteria population are happy and healthy all of the time and it will give you plenty of warning if you are doing something wrong on the system.

There are a few different ways that the bacteria inside the digester could be made sick or murdered by you, other people or even stray gases. It is down to you to think before you act in order to ensure that you do not cause any problems. If something has happened to the bacteria you need to work out what this was so that it never gets repeated again. You are the creator of the bacteria civilisation that is living in your garden it is down to you to ensure that you care for and protect this tiny metropolis that is in your care.

Don't worry too much about your bacteria suddenly dying on you as this shouldn't be a problem if you look after your digester properly. Generally speaking you should have a healthy and happy population of bacteria for many years to come.

Chapter 15

More books by this author

Solar & 12 Volt Power for Beginners

A complete step by step guide to creating your own off grid 12 volt solar power source written from the complete beginner's point of view.

- No technical jargon
- Simple instructions
- Full colour photos and diagrams
- Power wherever & whenever you want it
- Power laptops, TV's, power tools etc

Available at Amazon now!

Gasifiers Wood Gasification & Off Grid Power- A beginners guide

All the basic details needed for you to understand how gasifiers work without the technical jargon that confuses most people. Written especially for the beginner this book holds all the necessary information to get you started in the world of wood gasification.

- Free electricity
- Run vehicles free of charge
- Free hot water
- Free heating

Available at Amazon now!

Fireless Cookers Haybox Cookers & Retained Heat Cookers

Learn how to construct & use fantastic off grid cooking devices that can cook food using no energy at all. Lower your bills and make your life so much easier.

- Reduce cooking bills
- Reduce your carbon footprint
- Save money
- Save time
- Save the planet

Available at Amazon now!

Buying A Used Car-A Beginners Guide

Learn the basic checks and hidden faults that you have to look for when buying a used vehicle. With just a little bit of knowledge you can make buying used vehicles a lot less risky & expensive.

- Find good vehicles
- Detecting faults on vehicles
- Test driving vehicles
- Avoiding stolen & damaged vehicles

Available at Amazon now!

Goodbye and good luck.

I wish you all the best on your trip into being self sufficient and environmentally conscious with what you do with your waste.

Printed in Great Britain
by Amazon